# Results and Problems in Cell Differentiation

*Series Editors:*
W. Hennig, L. Nover, U. Scheer

**27**

# Springer

*Berlin*
*Heidelberg*
*New York*
*Barcelona*
*Hong Kong*
*London*
*Milan*
*Paris*
*Singapore*
*Tokyo*

Heribert Hirt (Ed.)

# MAP Kinases
# in Plant Signal Transduction

With 26 Figures and 5 Tables

 Springer

Prof. Dr. Heribert Hirt
University of Vienna
Vienna Biocenter
Institute of Microbiology and Genetics
Dr. Bohrgasse 9
A-1030 Vienna

ISSN 0080-1844
ISBN 3-540-65625-1  Springer-Verlag Berlin Heidelberg New York

Library of Congress Cataloging-in-Publication Data.
MAP kinases in plant signal transduction / Heribert Hirt, ed. p. cm. – (Results and problems in cell differentiation ; 27) Includes bibliographical references and index. ISBN 3-540-65625-1 (hardcover) 1. Protein kinases.
2. Plant enzymes. 3. Plant cellular signal transduction. 4. Mitogens. I. Hirt, Heribert, 1956– . II. Series. QH607.R4
vol. 27 [QK898.P79] 571.8'35 s–dc21 [572'.7922]

© Springer-Verlag Berlin Heidelberg 2000
Printed in Germany

Production: PRO EDIT GmbH, D-69126 Heidelberg
Cover Concept: Meta Design, Berlin
Cover Production: design & production, D-69121 Heidelberg
Typesetting: Zechner, D-67346 Speyer
SPIN: 10681955     39/3136 – 5 4 3 2 1 0 – Printed on acid-free paper

# Contents

**MAP Kinases in Plant Signal Transduction**
Heribert Hirt

| | | |
|---|---|---|
| 1 | MAPK Modules Form Basic Units of Eukaryotic Signal Transduction . . . . . . . . . . . . . . . . . | 1 |
| 2 | Different MAPK Pathways Exist for Different Signals . . . . . . | 2 |
| 3 | What Are So Many Plant MAPK Genes For? . . . . . . . . . . . | 3 |
| 4 | Arabidopsis: Emerging MAPK Modules . . . . . . . . . . . . . | 4 |
| 5 | MAPKs and Stress Signaling . . . . . . . . . . . . . . . . . . | 4 |
| 5.1 | Mechanical Stress . . . . . . . . . . . . . . . . . . . . . . . | 4 |
| 5.2 | Osmotic Stress . . . . . . . . . . . . . . . . . . . . . . . . . | 5 |
| 5.3 | Wounding . . . . . . . . . . . . . . . . . . . . . . . . . . . . | 5 |
| 5.4 | Plant-Pathogen Interaction . . . . . . . . . . . . . . . . . . . | 6 |
| 6 | Cell Cycle . . . . . . . . . . . . . . . . . . . . . . . . . . . . | 7 |
| 7 | Auxin . . . . . . . . . . . . . . . . . . . . . . . . . . . . . . | 8 |
| 8 | Abscisic Acid and Giberellins . . . . . . . . . . . . . . . . . . | 8 |
| 9 | Ethylene . . . . . . . . . . . . . . . . . . . . . . . . . . . . . | 8 |
| 10 | Concluding Remarks . . . . . . . . . . . . . . . . . . . . . . | 9 |

**MAP Kinases in Plant Signal Transduction: How Many, and What For?**
Wilco Ligterink

| | | |
|---|---|---|
| 1 | Introduction . . . . . . . . . . . . . . . . . . . . . . . . . . | 11 |
| 2 | The Structure of MAPKs . . . . . . . . . . . . . . . . . . . . | 12 |
| 3 | MAPKs in Yeast and Animals . . . . . . . . . . . . . . . . . . | 14 |
| 4 | Plant MAPKs . . . . . . . . . . . . . . . . . . . . . . . . . . | 14 |
| 5 | Plant MAPKs Can Be Classified into Five Groups . . . . . . . . | 18 |
| 6 | PERK Groups: Towards a Functional Classification System of Plant MAPKs . . . . . . . . . . . . . . . . . . . . . . . . . | 19 |
| 7 | How Many Are Out There? Plant ESTs with Similarities to MAPKs . . . . . . . . . . . . . . . . . . . . | 20 |
| | References . . . . . . . . . . . . . . . . . . . . . . . . . . . | 25 |

**MAP Kinase Cascades in *Arabidopsis*:**
**Their Roles in Stress and Hormone Responses**
Tsuyoshi Mizoguchi, Kazuya Ichimura, Riichiro Yoshida,
and Kazuo Shinozaki

1    Introduction . . . . . . . . . . . . . . . . . . . . . . . . . . . . . .    29
2    MAPK Cascades Involved in Stress Responses
     in Animals and Budding Yeast  . . . . . . . . . . . . . . . . . . .    30
3    *Arabidopsis* Homologs of Protein Kinases in MAPK Cascades  .    31
4    Transcriptional Control of the Putative Components
     of *Arabidopsis* MAPK Cascades in Response
     to Environmental Stresses . . . . . . . . . . . . . . . . . . . . . .    32
5    Activation of MAPK-Like Activity
     by Environmental Stresses in *Arabidopsis thaliana* . . . . . . . .    33
6    Analysis of Functional Interactions
     of the Components of MAPK Cascades
     in *Arabidopsis* Using the Yeast System . . . . . . . . . . . . . .    34
7    Putative Upstream Factors of MAPK Cascades
     in *Arabidopsis*  . . . . . . . . . . . . . . . . . . . . . . . . . . .    35
     References  . . . . . . . . . . . . . . . . . . . . . . . . . . . . . .    36

**MAP Kinases in Pollen**
Cathal Wilson and Erwin Heberle-Bors

1    Pollen Development and Germination . . . . . . . . . . . . . . .    39
2    Signal Transduction in Pollen . . . . . . . . . . . . . . . . . . . .    40
3    Changes in Pollen Grains After Hydration . . . . . . . . . . . .    42
4    MAP Kinases in Pollen  . . . . . . . . . . . . . . . . . . . . . . .    43
5    The Response of MAP Kinase Signaling Pathways
     to Osmotic Stress  . . . . . . . . . . . . . . . . . . . . . . . . . .    44
6    Specificity of the MAP Kinase Response . . . . . . . . . . . . .    46
     References  . . . . . . . . . . . . . . . . . . . . . . . . . . . . . .    48

**Mitogen-Activated Protein Kinases and Wound Stress**
Shigemi Seo and Yuko Ohashi

1    Introduction  . . . . . . . . . . . . . . . . . . . . . . . . . . . . .    53
2    Evidence for Activation of MAPK-Like Protein Kinases
     by Wounding . . . . . . . . . . . . . . . . . . . . . . . . . . . . .    54
3    Evidence for Activation of MAPKs by Wounding . . . . . . . .    55
4    Transcriptional Activation of MAPKs by Wounding  . . . . . .    56
5    Systemic Activation of MAPKs by Wounding . . . . . . . . . .    57

6     What Is the Initial Wound Signal for Activation of MAPKs?  . . .      58
7     Studies of Function of MAPKs Using Transgenic Plants  . . . . .       58
8     Concluding Remarks  . . . . . . . . . . . . . . . . . . . . . . . .    60
      References  . . . . . . . . . . . . . . . . . . . . . . . . . . . .    60

**Pathogen-Induced MAP Kinases in Tobacco**
Shuqun Zhang and Daniel F. Klessig

1     Introduction  . . . . . . . . . . . . . . . . . . . . . . . . . . .     66
2     Salicylic Acid-Induced Protein Kinase (SIPK) . . . . . . . . . . .      68
3     MAP Kinases Activated by a Cell Wall-Derived
      Carbohydrate Elicitor and Purified Proteinaceous Elicitins
      from *Phytophthora* spp. . . . . . . . . . . . . . . . . . . . .        69
3.1   SIPK . . . . . . . . . . . . . . . . . . . . . . . . . . . . . . .      69
3.2   Wounding-Induced Protein Kinase (WIPK) . . . . . . . . . . .           70
4     Activation of SIPK by Bacterial Harpin  . . . . . . . . . . . . .      72
5     Activation of SIPK and WIPK by TMV  . . . . . . . . . . . . .          72
6     Activation of SIPK and WIPK by Avr9
      from *Cladosporium fulvum* . . . . . . . . . . . . . . . . . . . .      74
7     Wounding and MAP Kinase Activation  . . . . . . . . . . . . .          76
8     General Discussion  . . . . . . . . . . . . . . . . . . . . . . .      78
      References  . . . . . . . . . . . . . . . . . . . . . . . . . . . .    81

**Receptor-Mediated MAP Kinase Activation in Plant Defense**
Heribert Hirt and Dierk Scheel

1     Introduction  . . . . . . . . . . . . . . . . . . . . . . . . . . .     85
2     *Phytophthora sojae* Elicitor-Induced Responses of Parsley Cells:
      A Model System for Plant-Pathogen Interaction  . . . . . . . . .       86
3     Pep-13 Elicitor Binds to a 100 kDa Plasma Membrane Receptor            87
4     Signal Transduction Events
      of Pep-13 Elicitor-Induced Defense Responses  . . . . . . . . . .      87
5     Pep-13 Elicitor Induces Activation of a MAP Kinase  . . . . . . .      88
6     MAP Kinase Activation Occurs Through the Pep-13 Receptor  .            89
7     MAP Kinase Activation Is Dependent on Ion Fluxes  . . . . . . .        89
8     MAP Kinase Activation Is Independent of Oxidative Burst  . . .         91
9     Activation of MAP Kinase Is Correlated
      with Its Nuclear Translocation  . . . . . . . . . . . . . . . . . .    91
      References  . . . . . . . . . . . . . . . . . . . . . . . . . . . .    92

## Regulation of Cell Division and the Cytoskeleton
## by Mitogen-Activated Protein Kinases in Higher Plants
László Bögre, Ornella Calderini, Irute Merskiene, and Pavla Binarova

| | | |
|---|---|---|
| 1 | Regulation of the $G_1$/S Transition: The Animal Paradigm . . . . | 95 |
| 2 | The Plant Players of $G_1$ Control . . . . . . . . . . . . . . . . . | 97 |
| 3 | Regulation of $G_2$/M Transition: The Animal Paradigm . . . . . . | 100 |
| 4 | Plant Mitosis; Mechanically Different | |
| | but the Regulation Might be the Same . . . . . . . . . . . . . . | 102 |
| 5 | Spatial Regulation of Cell Division and Growth | |
| | are Dictated by Plant Specific Cytoskeletal Structures . . . . . . | 105 |
| 5.1 | The Cortical Microtubules . . . . . . . . . . . . . . . . . . . . | 106 |
| 5.2 | The Preprophase Band . . . . . . . . . . . . . . . . . . . . . . | 107 |
| 5.3 | The Mitotic Spindle . . . . . . . . . . . . . . . . . . . . . . . | 108 |
| 5.4 | The Phragmoplast . . . . . . . . . . . . . . . . . . . . . . . . | 110 |
| | References . . . . . . . . . . . . . . . . . . . . . . . . . . . . | 111 |

## The MAP Kinase Cascade
## That Includes MAPKKK-Related Protein Kinase NPK1 Controls
## a Mitotic Process in Plant Cells
Ryuichi Nishihama and Yasunori Machida

| | | |
|---|---|---|
| 1 | Isolation of the NPK1 cDNA . . . . . . . . . . . . . . . . . . . | 119 |
| 2 | Identification of NPK1 as a MAPKKK . . . . . . . . . . . . . . | 120 |
| 3 | Structural Features of NPK1 | |
| | and Its *Arabidopsis* Homologs, ANP1/2/3 . . . . . . . . . . . . | 121 |
| 4 | Activity of ANP1 May be Regulated by Differential Splicing . . . | 124 |
| 5 | Expression Pattern of the NPK1 Gene . . . . . . . . . . . . . . | 124 |
| 6 | Upstream and Downstream Factors of NPK1 . . . . . . . . . . . | 126 |
| 7 | Possible Function of NPK1 . . . . . . . . . . . . . . . . . . . . | 128 |
| | References . . . . . . . . . . . . . . . . . . . . . . . . . . . . | 128 |

## Mitogen-Activated Protein Kinase
## and Abscisic Acid Signal Transduction
Sjoukje Heimovaara-Dijkstra, Christa Testerink, and Mei Wang

| | | |
|---|---|---|
| 1 | Introduction: Facts and Speculation . . . . . . . . . . . . . . . | 131 |
| 2 | Physiology of ABA . . . . . . . . . . . . . . . . . . . . . . . . | 132 |
| 2.1 | Physiological Function of ABA . . . . . . . . . . . . . . . . . . | 132 |
| 2.2 | Analogies Between ABA Function | |
| | and MAP Kinase Action in Plant, Yeast | |
| | and Vertebrate Physiology . . . . . . . . . . . . . . . . . . . . | 133 |

3       Signal Transduction Cascades of ABA . . . . . . . . . . . . . .   133
3.1     Tissue Specificity of ABA Action . . . . . . . . . . . . . . . .   133
3.2     Phosphorylation/Dephosphorylation Events . . . . . . . . . .      135
4       MAP Kinase Activation by ABA . . . . . . . . . . . . . . . . .     135
5       The Importance of Protein Phosphatases for ABA Signalling . .     138
5.1     Disruption of PP2C Genes Leads to ABA Insensitivity . . . . . .   138
5.2     Abi1 and Abi2 Phosphatases are Homologous
        to a Regulator of MAP Kinase Pathways in Yeast and Alfalfa . .    139
5.3     Where Do the Type 2C Phopshatases Fit In? . . . . . . . . . . .   139
6       Conclusions . . . . . . . . . . . . . . . . . . . . . . . . . .   140
        References . . . . . . . . . . . . . . . . . . . . . . . . . .    141

**Signal Transduction of Ethylene Perception**
Sigal Savaldi-Goldstein and Robert Fluhr

1       Introduction . . . . . . . . . . . . . . . . . . . . . . . . . .  145
2       Detection of Phosphorylation Events
        in Ethylene Signal Transduction . . . . . . . . . . . . . . . .   146
3       A Survey of Putative Genetic and Molecular Components
        of the Ethylene Signal Pathway . . . . . . . . . . . . . . . . .  147
3.1     Mutants of Ethylene Response . . . . . . . . . . . . . . . . . .  147
3.2     The ETR1 Protein and Related Homologues . . . . . . . . . . .     148
3.3     Functional Aspects of ETR1 Related Proteins . . . . . . . . . .   149
3.4     Response Regulator Homologues in *Arabidopsis* (ARRs) . . . .     151
3.5     CTR1: A Raf-Like Protein Kinase and Candidate Entry
        to MAP Kinases . . . . . . . . . . . . . . . . . . . . . . . . .  151
3.6     EIN3: A Nuclear-Localized Component of Ethylene Signaling . .     152
3.7     PK12: A Kinase Exhibiting Ethylene-Dependent Activity . . . .     153
3.8     EREBPs: Transcription Factors
        That Bind cis-Regulatory Sequences . . . . . . . . . . . . . . .  154
4       Dynamics of Ethylene Perception and Signal Transduction . . .     154
4.1     Perception of Ethylene and Phospho-Transfer . . . . . . . . . .   154
4.2     Linking Phospho-Relay and Kinase Cascades . . . . . . . . . . .   155
4.3     Redundancy of Components and Harmonizing
        the Ethylene Response . . . . . . . . . . . . . . . . . . . . .   156
        References . . . . . . . . . . . . . . . . . . . . . . . . . .    158

**Subject Index** . . . . . . . . . . . . . . . . . . . . . . . . . . .  163

# MAP Kinases in Plant Signal Transduction

Heribert Hirt

**Summary:** Mitogen-activated protein kinase (MAPK) pathways are modules involved in the transduction of extracellular signals to intracellular targets in all eukaryotes. Distinct MAPK pathways are regulated by different extracellular stimuli and are implicated in a wide variety of biological processes. In plants there is evidence for MAPKs playing a role in the signaling of abiotic stresses, pathogens, plant hormones, and cell cycle cues. The large number and divergence of plant MAPKs indicates that this ancient mechanism of bioinformatics is extensively used in plants and their study promises to give molecular answers to old questions.

## 1
## MAPK Modules Form Basic Units of Eukaryotic Signal Transduction

Mtogen-activated protein kinases (MAPKs) are encoded by a large family of serine/threonine protein kinases that are found in all eukaryotes. Activation of MAPKs is brought about by upstream MAPK kinases (denoted as MAPKKs) through phosphorylation of the conserved threonine and tyrosine residues (Fig. 1). A given dual specificity MAPKK can only activate a specific MAPK and cannot functionally substitute other MAPKKs. MAPKKs are themselves activated by phosphorylation through upstream kinases that belong to the class of MAPKK kinases (MAPKKKs), including the Raf and Mos kinases. A specific set of three functionally interlinked protein kinases (MAPKKK-MAPKK-MAPK) forms the basic module of a MAPK pathway (Fig. 1). MAPK pathways may integrate a variety of upstream signals through interaction with other kinases or G proteins. The latter factors often serve directly as coupling agents between plasma membrane located sensors of extracellular stimuli and cytoplasmic MAPK modules.

---

[1] Institute of Microbiology and Genetics, Vienna Biocenter, Dr. Bohrgasse 9,
1030 Vienna, Austria

Results and Problems in Cell Differentiation, Vol. 27
Hirt (Ed.): MAP Kinases in Plant Signal Transduction
© Springer-Verlag Berlin Heidelberg 2000

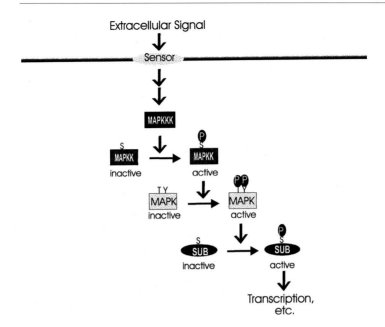

**Fig. 1.** Simplified scheme for MAP kinase signal transduction cascades. An extracellular signal is received by a membrane-located receptor. The activation of the MAP kinase module (MAPKKK activates MAPKK, which activates MAPK) by the receptor may occur via several intermediate steps and by different routes. The active MAP kinase may activate other protein kinases, phosphorylate cytoskeletal components or translocate to the nucleus and activate transcription factors giving rise to expression of specific genes. Arrows indicate activation

At the downstream end of the module, activation of the cytoplasmic MAPK module often induces the translocation of the MAPK into the nucleus where the kinase activates certain sets of genes through phosphorylation of specific transcription factors (Fig. 1). In other cases, a given MAPK may translocate to other sites in the cytoplasm to phosphorylate specific enzymes (protein kinases, phosphatases, lipases, etc.) or cytoskeletal components. By tight regulation of MAPK localisation and through expression of certain signaling components and substrates in particular cells, tissues, or organs, particular MAPK pathways can mediate signaling of a multitude of extracellular stimuli and bring about a large variety of specific responses.

## 2
## Different MAPK Pathways Exist for Different Signals

MAP kinase pathways are best understood in yeast and this organism can be viewed as a model organism for understanding the role of MAPK cascades in more complex multicellular systems. Of the six MAPK genes that are present in the yeast genome, functions for five MAPKs have been identified. The FUS3 and KSS1 MAPKs are activated by exposure of yeast to pheromone and mutants of this pathway are sterile. Some of the components of the pheromone pathway, including the KSS1 kinase, are involved in pseudohyphal growth control in response to nitrogen starvation in diploid cells and invasive growth in haploid cells. Whereas the MPK1 MAPK is required for adaptation to a hypoosmolar environment, the HOG1 MAPK is activated under hyperosmolar conditions and results in increased biosynthesis of glycerol which acts as an osmotic stabilizer. The SMK1 MAPK is involved in the control of spore formation. Mainly from studies in yeast, we know that different MAPKs can not substitute for each other and that specific interactions with other proteins are brought about by scaffold proteins that serve as interaction platforms.

In mammals, MAPKs were originally identified as transducers of mitogens (proliferation-inducing substances). Later, MAPKs were also shown to be involved in signaling hormones, neurotransmitters, and signals for differentiation. This group of MAPKs was therefore called ERKs (for extracellular signal-regulated kinases). Two groups of protein kinases have been added to the family of mammalian MAPKs. The stress-activated protein kinases (SAPKs) or Jun kinases (JNKs) were identified by their ability to specifically phosphorylate the transcription factor c-jun, mediating transcription of specific genes following exposure to ultraviolet radiation, pro-inflammatory cytokines and environmental stress. The second family, the p38 kinases are activated in response to endotoxin from gram-negative bacteria, interleukin-1, and hyperosmolar stress. A p38 kinase is also induced by heat shock, activating yet another protein kinase that phosphorylates small heat shock proteins. The JNK and p38 kinases not only share functional similarities in terms of stress signaling, but can also complement HOG1-deficient mutants of yeast.

## 3
## What Are So Many Plant MAPK Genes For?

By searching public databases Ligterink (this vol.) has found a total of 23 MAPK genes from several plant species. The predicted amino acid sequences show high conservation to animal and yeast MAPKs with highest similarity in the eleven domains that are necessary for the catalytic function of serine/threonine protein kinases. The threonine and tyrosine residues whose phosphorylation is necessary for activation of MAPKs in all eukaryotes are also found in all plant MAPKs. Although the sequences outside the 11 kinase domains show little homology to other MAPK sequences from the same species, these motifs are highly conserved in specific MAPKs of other plant species and form the basis for classification into five subfamilies. MAPKs within one subfamily appear to perform similar functions.

Analysis of MAPK sequences encoded by partial cDNAs and expressed sequence tag libraries indicates the existence of at least one additional subfamily that shows considerable divergence to other MAPKs. According to this analysis, it is clear that plants contain an unexpectedly large and divergent family of MAPKs, but functional information is available for only a small subset of these sequences.

# 4
# Arabidopsis: Emerging MAPK Modules

Considering the importance of Arabidopsis as a model organism for the plant biology field, Mizoguchi et al. provide a review of the current knowledge on Arabidopsis MAPKs and their potential upstream regulators (see Mizoguchi et al., this vol.). The authors show that one approach to connect the bewildering number of MAPKKKs, MAPKKs, and MAPKs into distinct MAPK modules, may be by yeast two-hybrid interaction analysis and other yeast techniques.

# 5
# MAPKs and Stress Signaling

Due to their sessile habit, plants are exposed to a variety of environmental stresses, including pathogens as well as changes in temperature, water conditions, and wind. MAPKs have been implied in signal transduction of all of these stresses.

## 5.1
## Mechanical Stress

Wind is a mechanical stress and can lead to major changes in the growth pattern of plants, diverting energy into strengthening the plants architecture nicely seen in the short stature of wind-exposed trees in the mountains or in coastal areas. Experiments showing that mechanical manipulation of Arabidopsis leaves induces transcription of particular MAPK and MAPKK genes suggest that a MAPK pathway might signal mechanical stimuli. Direct proof was obtained when it was shown that touching alfalfa leaves for 2 s is sufficient to induce a transient activation of a MAPK. Whereas constantly shaken suspension cultured cells were found to have constitutive levels of activated MAPK, this kinase activity vanished when cells were allowed to rest for 1 h, but shaking for a few seconds restored activity.

## 5.2
## Osmotic Stress

Osmotic stress, originating from water shortage, high soil salinity, or freezing, is one of the most severe limiting factors for the development and growth of plants. Plants in different climatic zones have developed specific mechanisms to withstand these stresses. The adaptive strategies mainly depend on the expression of specific sets of genes that result in changes in the composition of the major cell components. A MAPK has been shown to be involved in the transmission of drought and cold signals in alfalfa. Assuming induction of MAPK gene expression by a particular stimulus as evidence for a role in signal transduction, it is likely that MAPK pathways also play a role in response to other physical stresses, because transcript amounts of specific Arabidopsis MAPK and MAPKK genes increased upon water stress, cold, touch and high salt.

Ntf4, a tobacco MAP kinase, was recently shown to be expressed and activated in pollen. Although both the expression and the activity of Ntf4 are developmentally controlled during pollen maturation, hydration of the mature dry pollen can stimulate the activity of Ntf4 much further. Wilson and Heberle-Bors discuss the present evidence for a role of Ntf4 in pollen development and of MAPKs in osmotic stress in general (Wilson and Heberle-Bors, this vol.).

## 5.3
## Wounding

In addition to abiotic challenges, plants come into contact with a variety of potential pathogens. Survival of plants in this environment is only possible through sophisticated defense and adaptation strategies, and it comes as no surprise to find MAPKs to be involved in signal transduction of pathogen attack.

Wounding of plant tissues may be caused by mechanical injury, pathogen, or herbivore attack. To counter this challenge, plants have developed defense systems that are mostly based on activation of particular sets of genes encoding a variety of enzymes, pathogen response (PR) proteins or proteinase inhibitors. Whereas some of these genes are only induced locally at the site of attack, others are expressed systemically throughout the plant and protect the plant from attack at distant sites. Several of the genes involved in defense response have been identified and studied, but relatively little is known about how a plant senses wounding and transmits the signal to the nucleus before induction of the respective defense response genes. Seo and Ohashi were among the first to show that a specific MAPK was involved in wound signal transduction in tobacco and review the present status in this field (Seo and Ohashi, this vol.). Because a MAPK is activated by wounding within less than 1 min in both monocots and dicots, MAPKs are placed in the very first line of responses to a wound signal. Wounding of tobacco leaves was shown to lead to the rapid accumulation of transcripts of a particular MAPK gene, termed WIPK for wound-induced protein kinase. Overexpression of the WIPK gene in transgenic tobacco leads to suppression of the wound response, including the wound-inducible increase of jasmonic acid and its methyl ester, substances that are normally induced by wounding and that are involved in mediating the systemic wound response.

Insight into the role of a MAPK in wound signaling was provided by the finding that systemin, an 18-amino acid peptide that confers a systemic wound response to tomato, also activates a MAPK. Two other substances that are released during wounding by pathogens, chitosan and polygalacturonic acid, were able to induce a similar kinase activation pattern. As shown by the *def1* mutant that is defective in the jasmonic acid pathway, wound defense gene signaling in tomato is mediated by jasmonic acid. The jasmonic acid pathway is not necessary for wound-induced kinase activation because *def1* mutant tomato plants still showed MBP kinase activation in response to systemin, chitosan, and polygalactuonic acid, but not to jasmonic acid. These results are consistent with the model that a MAPK is involved in translating the wound signal into the synthesis of jasmonic acid.

In all investigated plant species, the activation of the wound-induced MAPKs is a post-translational process. However, the inactivation is dependent on de novo transcription and translation of protein factors. One of these newly synthesized proteins was recently identified as MP2C, a protein phosphatase of type 2C. MP2C is part of a negative feedback loop, whereby MP2C expression is tightly regulated through the activity of the wound-activated MAPK. By inactivating the MAPK pathway, the phosphatase shuts down its own synthesis and thereby helps to reset the pathway. This feedback loop provides the cell with a mechanism for using the components of a signaling pathway repeatedly for monitoring whether more pathogens are lining up in front of the city walls.

## 5.4
## Plant-Pathogen Interaction

Zhang and Klessig address the recent advances implicating MAPKs in plant-pathogen signal transduction (Zhang and Klessig, this vol.). While plants may sense the presence of some pathogens through wounding, others are recognized by other means, involving cell wall components and microbial elicitors. Treating tobacco leaves with harpins, protein elicitors from the bacterial pathogen *Erwinia amylovora*, or tobacco cells with elicitors, derived from fungal cell walls of *Phytophthora species,* results in activation of protein kinases. Although these kinases have similarities with MAPKs, only one of them was unequivocally identified as SIPK. WIPK and SIPK have recently been shown to be activated by infection of tobacco plants with TMV (tobacco mosaic virus), suggesting that chemically distinct factors derived from bacteria, fungi, and viruses can activate MAPK pathways in plants.

In another established plant-pathogen model system, that between the fungus *Phythophthora sojae* and the non-host parsley, evidence was also provided that a specific MAPK is part of the information transfer process from recognition of the pathogen to defense gene activation. In this system, extensive studies have charted a presumptive signal transduction pathway, whereby a peptide elicitor derived from a secreted glycoprotein of the fungal pathogen binds to a plasma membrane-located receptor and sequentially activates various ion channels, an NADPH oxidase (resulting in an oxidative burst), and induction of PR genes and phytoalexin synthesis. Treatment of cells with the peptide elicitor activates an elicitor-regulated MAP kinase. Hirt and Scheel discuss the current evidence (Hirt and Scheel, this vol.) indicating that within this pathway, the MAPK acts downstream of the receptor and the elicitor-activated ion channels but upstream or independently of the NADPH oxidase. Interestingly, after treating parsley cells with elicitor, the MAPK becomes rapidly translocated from the cytoplasm into the nucleus, suggesting a direct role of thus MAPK in the regulation of elicitor-induced gene transcription.

# 6
# Cell Cycle

Historically, the name MAPK was derived from the fact that mitogens activated the ERK1 and ERK2 protein kinases in mammalian cells. Bögre et al. (this vol.) review the facts that MAPKs function as cell cycle regulators not only in animals but also in plants. The authors also discuss the possibility that MAPKs might provide a possible link between the cytoskeleton and the cell cycle machinery.

So far, a clear role for MAPKs in plant cell cycle regulation has been provided in two species. The closely related alfalfa MMK3 and tobacco Ntf6 MAPKs were found to become activated during late mitosis. The activation of the MAPKs during the anaphase to telophase transition was correlated with a transient localisation of the kinases at the cell plate. These results suggest that these MAPKs might play a role in cytokinesis.

Nishihama and Machida (this vol.) give additional support for this hypothesis, reporting that NPK1, a tobacco MAPKKK, might function in mitosis. Interestingly, NACK1, an interacting partner of NPK1, appears to be a microtubule-associated motor protein. The similar transient cell cycle phase-specific localisation of NACK1, NPK1, and the Ntf6 and MMK3 MAPKs to the cell plate, suggests that these proteins might be part of the same MAPK module with a role in phragmoplast assembly or cell plate formation.

# 7
# Auxin

Most plant cell cultures require auxin for proliferation, showing that auxin may act as a mitogen under certain conditions. Whereas auxin starvation arrests cell division in a tobacco cell suspension culture, readdition starts the cell cycle. During this process, a MAPKK and a protein kinase that has the properties of a MAPK are activated, suggesting that a MAPK pathway is involved in signal transduction of auxin.

Recently, expression of truncated NPK1 was found to negatively affect expression of auxin-responsive genes. It is presently unclear whether the effects of NPK1 on auxin-dependent gene expression and its putative role in cytokinesis are related.

# 8
# Abscisic Acid and Giberellins

Abscisic acid (ABA) influences many processes in plant physiology, such as embryo development, seed germination and abiotic stress responses, including adaptation to drought and salt stress. Besides giving an overview of ABA signal transduction, Heimovaara-Dijkstra et al. (this vol.) also discuss the evidence that ABA can induce a MAPK. Treating barley aleurone protoplasts with ABA, a MAPK-like activity was found to be induced. Besides inducing specific genes, ABA is known to inhibit giberellic acid (GA)-dependent effects in aleurone cells, such as the expression of hydrolytic enzymes, stimulation of protein synthesis, and breakdown of storage reserves. Whereas ABA stimulates MAPK activation in aleurone cells, GA may do the reverse, as indicated by the negative effect of GA on transcript accumulation of a MAPK gene in oat aleurone cells.

# 9
# Ethylene

Savaldi-Goldstein and Fluhr give an overview of how ethylene is perceived and signaled by plant cells (this vol.). Genetic analysis of the ethylene pathway in Arabidopsis indicates a possible involvement of a MAP kinase module. A number of mutants have been isolated that show a constitutive triple response (CTR) in the absence of ethylene. The *ctr1* mutant was isolated and the affected gene cloned and analyzed. The encoded CTR1 protein was found to be similar to mammalian Raf kinase. Upstream components in the CTR1 pathway appear to be the plasma membrane located ethylene receptors ETR1 and ERS, encoding proteins with homology to two-component sensor regulator proteins that are well-known signaling transducers in bacteria. Bacterial two-component systems are responsible for regulating a variety of processes, including chemotaxis, osmolarity and nutrient availability. The sensor is usually a membrane located kinase that autophosphorylates on a histidine residue upon receiving the extracellular signal. The regulator then becomes phosphorylated on an aspartate residue obtaining the phosphate group from the histidine of the sensor. The regulator can either be a separate protein or part of a hybrid sensor-regulator kinase. The ETR1 protein is a hybrid sensor-regulator kinase and resembles the SLN1 protein that is responsible for signaling hyperosmotic stress via the HOG1 MAPK pathway in yeast. Because two-component systems have an inbuilt short term inactivation mechanism, these signaling components are optimally suited to respond to rapid changes of extracellular signals, and may be found in many other eukaryotic pathways.

# 10
# Concluding Remarks

Over the last few years we have seen a surge of results claiming an involvement of MAP kinases in a variety of signaling processes in plants. In many cases, however, only indirect proof has been provided and the responsible protein kinase/gene has not been identified. Without these results, we should keep in mind that other types of protein kinases share many of the properties of MAP kinases (similar substrate specificities, molecular masses, etc.). Therefore, it will be essential to obtain the proper tools in order to identify the specific MAP kinases and isolate the respective genes encoding the enzymes. For these purposes, biochemical and genetic approaches will be essential and should contribute equally to assign specific functions to distinct MAPKs.

Assuming that the present evidence will hold, how is it that so many signals can be transmitted by MAPK pathways? From a theoretical standpoint, there are surely enough MAPK genes in the plant genome to assign each of these genes a specific role. However, experience from a growing number of investigations of mammalian cells tells us that things might be much more complicated. This is shown by the fact that a given extracellular signal usually activates not just a single, but several independent pathways. A single pathway may also be activated by a number of other unrelated signals. The same signal may activate different pathways in different cells. Last but not least, the activation of a pathway is not an all or nothing process, but can be a transient or constitutive event and may differ in amplitude. Changing only one of the above parameters has been shown to dramatically affect the outcome of the cellular response. Contemplating along these lines, signal transduction is likely to be a question of pathway combinations, and the responses at the chromatin level may depend on which of the many protein kinase pathways are active, and to what extent, at a given moment.

# MAP Kinases in Plant Signal Transduction: How Many, and What For?

Wilco Ligterink

**Summary:** Mitogen-activated protein kinase (MAPK) pathways are protein kinase cascades that have a function in the transduction of extracellular signals to intracellular targets in all eukaryotes. Distinct MAPK pathways are regulated by different signals and have a role in a wide variety of physiological processes. In plants there is evidence for a role of MAPKs in the signaling of pathogens, abiotic stresses, plant hormones, and cell cycle cues. A large number of distinct MAPKs in plants have been identified that are all most similar to the animal ERK MAPKs. By sequence alignment all available full length plant MAPKs can be grouped into five subfamilies. Functional data exist for members of four subfamilies and show that different subfamilies encode MAPKs for specific functions. Analysis of partial MAPK sequences from full length, truncated cDNAs and expressed sequence tags (ESTs) revealed the presence of two new subfamilies in the plant MAPK superfamily. Signature sequences valid for the superfamily of plant MAPKs and each subfamily were derived and should help in future classification of novel MAPKs. The future challenge is to unambiguously assign functions to each MAPK and decipher the other partners of their signaling pathways.

# 1
# Introduction

Mitogen-activated protein kinases (MAPK) are serine/threonine protein kinases that can be found in all eukaryotes. The MAPK family belongs to the eukaryotic protein kinase superfamily. Within this superfamily they group together with the cyclin-dependent kinases (cdk), glycogen synthase kinases 3 (gsk3), and cell division cycle-dependent (cdc)-like kinases (Hanks and Hunter 1995).

[1] Institute of Microbiology and Genetics, Vienna Biocenter, Dr. Bohrgasse 9,
    1030 Vienna, Austria

Results and Problems in Cell Differentiation, Vol. 27
Hirt (Ed.): MAP Kinases in Plant Signal Transduction
© Springer-Verlag Berlin Heidelberg 2000

MAPKs have a function in transducing a broad set of extracellular signals to cellular targets and perform their function as part of a protein kinase cascade. Besides MAPKs, this cascade is formed by MAPK kinases (MAPKKs) and the MAPKK kinases (MAPKKKs). Signaling into the MAPK cascade can occur via tyrosine kinase receptors, G-protein-linked receptors and protein kinase C-dependent pathways. Downstream of the MAPK cascade the signal is mediated to cellular responses. MAPKs may be translocated to the nucleus to phosphorylate and thereby activate specific transcription factors or may stay in the cytoplasm and phosphorylate certain enzymes like protein kinases and phosphatases or cytoskeletal components.

MAPKs are activated through phosphorylation by the dual-specificity MAPKKs also named MAPK/ERK kinases (MEKs), on both the tyrosine and threonine residues in the conserved TXY motif (Marshall 1994). Phosphorylation on both residues is required for full activation of the MAPKs. Inactivation of MAPKs occurs through dephosphorylation by serine/threonine phosphatases, tyrosine phosphatases or dual specificity phosphatases (Canagarajah et al. 1997).

MAPKKs are activated through phosphorylation by MAPKKKs, also named MEK kinases (MEKKs), on serine and threonine/serine residues in the conserved SXXXS/T motif. MAPK cascades have been identified in yeast and animals. Recently the first possible MAPK cascade in plants has been identified (Mizoguchi et al. 1998).

Specificity of MAPK pathways is obtained by the activation through distinct MAPKKs and by the substrate specificity of the MAPKs. Much of the specificity of MAPK pathways is also obtained by physical separation of MAPK modules. Scaffold proteins have been identified that hold specific components together. Furthermore, particular MAPK pathways can mediate different extracellular signals to a broad range of responses by differential expression of certain signal pathway components and substrates in specific cells.

## 2
## The Structure of MAPKs

Sequence comparison was used to define conserved sequence motifs, establishing 11 subdomains that are present in all serine/threonine protein kinases (Hanks et al. 1988). Crystallographic analysis of ERK2 in the inactive as well as in the dual phosphorylated active conformation determined the folding of these conserved subdomains and has defined the 3-D architecture (Fig. 1; Zhang et al. 1994; Canagarajah et al. 1997). ERK2 has a two-lobed structure, where the active site is found at the domain interface (Goldsmith and Cobb 1994). The N-terminal domain is mainly composed of β-strands, whereas the C-terminal domain consist mainly of α-helices. This overall topology of ERK2 is similar to that of cAMP-dependent protein kinase and CDK2 (Taylor and Radzio-Andzelm 1994).

**Fig. 1.**
Ribbon model of the three-dimensional structure of ERK2 in the inactive, unphosphorylated state. The image was drawn using Rasmol from Roger Sayle and the coordinates from rat ERK2 (PDB ID = 1ERK). Labels for the secondary structural elements are based on Zhang et al. (1994). β strands are drawn in *light grey* and α helices in *dark grey*

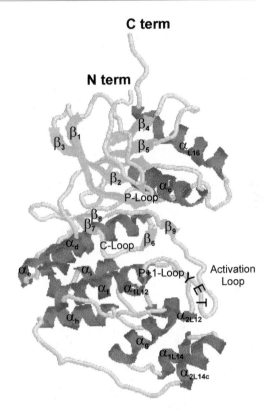

The loops between β-strands 6 and 7 and between β-strands 8 and 9 are important for forming the ATP binding pocket. The loop between β-strands 6 and 7 is called the catalytic loop (C-loop in Fig. 1) and contains the aspartate that functions as the catalytic base, the lysine that binds the γ-phosphate of ATP and the asparagine that may bind a metal ion. The loop between β-strands 8 and 9 contains the aspartate that chelates the primary $Mg^{2+}$ ion that bridges the α- and γ-phosphates of ATP. This loop may also have a function in stabilising contacts between the two lobes of the MAPK (Taylor and Radzio-Andzelm 1994).

Another important secondary structure element in the active site of MAPKs is the phosphate anchor loop ( P-loop in Fig. 1). The P-loop consists of the conserved sequence motif GXGXXG and helps to anchor the non-transferable phosphates of ATP.

MAPKs are only able to phosphorylate substrates that contain proline in the P+1 site. The P+1 site is the residue that immediately follows the serine or threonine residue of MAPK substrates. This substrate specificity is regulated by the P+1 loop. In ERK2, specific binding of substrate proline by this loop is possible after activation of the kinase.

One of the most important and least stable secondary structure elements of MAPKs is the activation loop containing the TXY motif (Fig. 1). The activation loop forms the mouth of the active site. The name of this loop is based on its importance for the activation of MAPKs. Both phosphorylation sites, T at position 183 and Y at position 185 in ERK2, are located in the activation loop (Zhang et al. 1994). In inactive ERK2, Y185 is not accessible and binding of the MEK to the MAPK already induces conformational changes so that both T183 and Y185 become accessible for phosphorylation. Comparison of the inactive and the activated ERK2 structures revealed strong conformational changes in the activation loop and in the L16 helix. These conformational changes align the catalytic residues and remodel the P+1 loop, resulting in an active kinase (Canagarajah et al. 1997).

# 3
# MAPKs in Yeast and Animals

Only six MAPKs are found in the fully sequenced yeast genome. Five of these MAPKs have been assigned to specific pathways: pheromone signaling, filamentous growth, adaptation to high osmolarity, cell wall remodeling, and sporulation (Madhani and Fink 1998).

Presently, there are 37 human MAPK protein sequences available in the database, but many of these sequences only differ in a single amino acid or are splicing variants, reducing the number to a total of 12. The ERK kinases are known to transmit a large array of signals to stimulate proliferation or differentiation. Members of the c-Jun amino-terminal kinases/stress-activated protein kinases (JNKs/SAPKs) and of the p38 kinases are involved in mediating various stresses.

# 4
# Plant MAPKs

Twenty-three full length MAPK cDNAs have been isolated from a variety of plant species (Table 1). Among these are nine *Arabidopsis*, five tobacco, and four alfalfa genes, and single genes from pea, parsley and petunia. There are two single representatives from monocot species, oat and wheat. The predicted amino acid sequences of these plant MAPKs show a highly conserved overall structure with sequence similarities between 50 and 98 % (Fig. 2). Highest similarity is observed in the 11 kinase subdomains (denoted by roman numerals in Fig. 2). As a reference, the animal ERK2 and p38 MAPKs are included in the alignment. Positions of $\beta$ strands ($\beta_{1-9}$), $\alpha$ helices ($\alpha_{c-i}$), and the connecting loops (L0–L16) as defined for ERK2 (Zhang et al. 1994) are also shown.

The entire MAPK family, including all yeast, animal and plant sequences, can be divided into three major subgroups: the stress-activated protein kinas-

**Table 1.** Summary of the isolated plant MAPK sequences

| Name | Species | Accession No. | Amino acids no. | MW (kDa) | pI (pH units) | Group/ subfamily | Reference |
|---|---|---|---|---|---|---|---|
| ATMPK1 | Arabidopsis thaliana | D14713 | 370 | 42.7 | 6.8 | 3/PERK5 | Mizoguchi et al. (1994) |
| ATMPK2 | Arabidopsis thaliana | D14714 | 376 | 43.1 | 6.1 | 3/PERK5 | Mizoguchi et al. (1994) |
| ATMPK3 | Arabidopsis thaliana | D21839 | 370 | 42.7 | 5.5 | 1/PERK2 | Mizoguchi et al. (1993) |
| ATMPK4 | Arabidopsis thaliana | D21840 | 376 | 42.9 | 5.7 | 2/PERK3 | Mizoguchi et al. (1993) |
| ATMPK5 | Arabidopsis thaliana | D21841 | 376 | 43.1 | 5.4 | 2/PERK3 | Mizoguchi et al. (1993) |
| ATMPK6 | Arabidopsis thaliana | D21842 | 395 | 45.1 | 5.2 | 1/PERK1 | Mizoguchi et al. (1993) |
| ATMPK7 | Arabidopsis thaliana | D21843 | 368 | 42.4 | 6.8 | 3/PERK5 | Mizoguchi et al. (1993) |
| ATMPK8 | Arabidopsis thaliana | unpublished | – | – | – | 4 | Mizoguchi et al. (1997) |
| ATMPK9 | Arabidopsis thaliana | unpublished | – | – | – | 4 | Mizoguchi et al. (1997) |
| NtF3 | Nicotiana tabacum | X69971 | 372 | 42.8 | 6.2 | 3/PERK5 | Wilson et al. (1993) |
| NtF4 | Nicotiana tabacum | X83880 | 393 | 45.1 | 5.5 | 1/PERK1 | Wilson et al. (1995) |
| NtF6 | Nicotiana tabacum | X83879 | 371 | 42.7 | 5.0 | 2/PERK4 | Wilson et al. (1995) |
| SIPK | Nicotiana tabacum | U94192 | 393 | 45.2 | 5.4 | 1/PERK1 | Zhang and Klessig DF (1997) |
| WIPK | Nicotiana tabacum | D61377 | 375 | 42.9 | 5.1 | 1/PERK2 | Seo et al. (1995) |
| MMK1 | Medicago sativa | X66469 | 387 | 44.4 | 5.4 | 1/PERK1 | Duerr et al. (1993) Jonak et al. (1993) |
| MMK2 | Medicago sativa | X82268 | 371 | 42.3 | 6.1 | 2/PERK3 | Jonak et al. (1995) |
| MMK3 | Medicago sativa | unpublished | 374 | 43.0 | 4.8 | 2/PERK4 | Bögre et al. (1999) |
| MMK4 | Medicago sativa | X82270 | 371 | 43.0 | 5.6 | 1/PERK2 | Jonak et al. (1996) |
| AsPK9 | Avena sativa | S56638 | 369 | 42.9 | 5.4 | 1? | Huttly and Phillips (1995) |
| ERMK | Petroselinum crispum | Y12785 | 371 | 42.8 | 5.7 | 1/PERK2 | Ligterink et al. (1997) |
| PhERK1 | Petunia x hybrida | X83440 | 384 | 44.4 | 6.8 | 3/PERK5 | Decroocq-Ferrant et al. (1995) |
| PsD5 | Pisum sativum | X70703 | 394 | 45.1 | 5.3 | 1/PERK1 | Stafstrom et al. (1993) |
| WCK1 | Triticum aestivum | AF079318 | 369 | 42.8 | 5.3 | 1? | Takezawa (1998) |

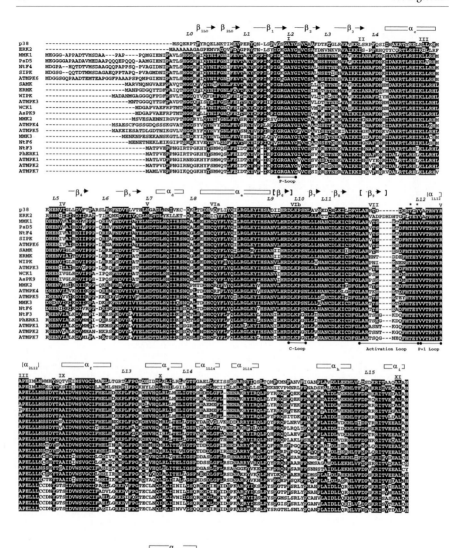

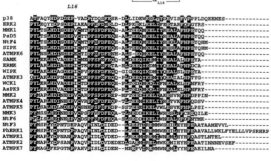

◄

**Fig. 2.** Multiple sequence alignment of the predicted amino acid sequences of 21 known plant MAPKs together with human p38 and ERK2 MAPKs. Amino acids are shown in the single letter code. Identical amino acids are shown in *white on black background*. The 11 kinase domains are indicated above the sequences with *roman numerals*. Secondary structure information is also placed above the sequences and is based on Wang et al. (1997) and Zhang et al. (1994). The phosphorylation sites are marked with *asterisks*

es (SAPKs), the extracellular signal-regulated kinases (ERKs) and the MAPK3 subgroup (Kültz 1998). Animal and fungal MAPK members can be found for all three subgroups. However, all plant MAPKs fall within the ERK subfamily and were therefore denoted as plant extracellular regulated protein kinases (PERK). Based on this phylogenetic analysis it was suggested that the ancestor of plant MAPKs split off from the animal/fungal lineage before the SAPK and MAPK3 subgroups evolved.

The presence of unique amino acid residues at the same positions in all members within a single subgroup can help to characterise new members. The PERK subfamily only has one unique residue if compared with the entire MAPK family (P39 in ATMPK3, Kültz 1998). This is an indication that the PERKs signal a broader range of responses than any of the individual subfamilies in animals, which are defined by a higher number of unique residues.

A striking sequence difference among members of ERK and SAPK subgroups was found in the activation loop. ERK2 has a six amino acid insertion in this loop compared with p38, a SAPK member. PERKs contain an insertion of only two to three amino acids at this position (Fig. 2). A characteristic motif in the activation loop is defined by the amino acid in between the T and Y MAPK phosphorylation sites. ERK2 has an E at this position (TEY) while p38 has a G (TGY). All plant MAPKs have a TEY motif. The function of the insertion in the activation loop and the TEY versus TGY motif was studied by site directed mutagenesis (Jiang et al. 1997). It was found that the dual phosphorylation motif and the insertion loop influence substrate specificity but are not crucial for the specificity of activation by upstream signals, while the length of the insertion loop plays a role in controlling autophosphorylation.

Besides the variable N- and C- terminal sequences found in plant MAPKs, short stretches of variable sequences are interspersed between the eleven conserved subdomains. These variable sequences occur mainly in the loops that join the secondary structural elements, and are typically found on the surface of the molecule in the ERK2 and p38 kinases. These variable parts might be important in determining the binding of a particular MAPK to upstream regulators or specific substrates. In chimeras constructed from p38 and ERK2, an N-terminal region of 40 residues, composed of the $\alpha_c$, L5, $\beta_4$, L6, $\beta_5$ structural units, was found to be involved in specifying the response of the kinase to different external signals, whereas the carboxy-terminal half of the molecule specified substrate recognition (Brunet and Pouysségur 1996).

# 5
# Plant MAPKs Can Be Classified into Five Groups

According to sequence similarity of predicted proteins, at least three different subgroups could be established within the plant MAPKs (Group 1–3) which could be further subdivided into at least five subfamilies, named PERK1–5 (Fig. 3 and Table 1). Only the two known MAPKs of monocots (AsPK9 and WCK1) could not be assigned unequivocally to any of these five subfamilies.

In addition to the phylogenetic analysis, a search was performed for unique residues characteristic for members within a group or subfamily. Due to the high sequence variability in group 1 MAPKs, no unique residues could be identified. This variability might reflect functional heterogeneity, suggesting that members of this group transmit a wide range of stimuli.

Members in the PERK1 subfamily have 13 unique residues: G3, D11, M14, A17, S41, I55, K71, V214, D260, M278, S285, R335 and A373 in ATMPK6. When compared with the other plant MAPKs, all members within the PERK1 subfamily contain a prolonged N-terminus (Fig. 2), resulting in higher calculated molecular masses (Table 1). Analysis of chimeras between ERK2 and p38 MAPKs indicated that the N-terminal region is involved in binding to upstream regulators (Brunet and Pouysségur 1996). The N-terminal extensions of the PERK1 subfamily might have a similar role. The deletion of one amino acid between subdomains X and XI was also recognised in all PERK1 subfamily members (Fig. 2).

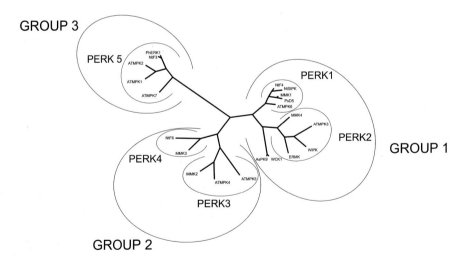

**Fig. 3.** Phylogenetic tree based on the amino acid sequences of all described 21 plant MAPKs (Table 1). This tree was reconstructed using the neighbor-joining method implemented in version 4.0.0d64 of PAUP* written by David L. Swofford. The three plant MAPK subgroups are denoted as GROUP 1–3, the five different subfamilies as PERK 1–5

Only four unique residues were found in members of the PERK2 subfamily (Q20, T58, L108, and D302 in ATMPK3). If the two monocot MAPKs, AsPK9 and WCK1, are also included in the PERK2 subfamily, there are five additional unique residues (M77, T119, N234, T259, and D270 in ATMPK3).

The PERK3 and PERK4 subfamilies are too small to define subfamily-specific unique residues. Therefore, unique residues are given only for the entire group 2. These are G18, G75, E220, Y249, S273, and V280 in ATMPK4.

A total of 55 unique residues were identified in the PERK5 subfamily (V5, N9, G10, G15, K16, H17, Y18, S20, M21, W22, G23, D29, K36, S49, L80, R87, M98, K107, L111, K125, F140, G162, Q188, C207, D209, N210, G211, T212, S213, L228, G229, T237, E238, C239, L240, N241, N248, Q253, P264, K265, S272, Y275, S276, G278, A288, Q297, S309, A331, I335, L337, D340, R349, Y359, H360, and A363 in ATMPK1). Another characteristic feature of the PERK5 subfamily is the shorter N-terminus and extended C-terminus when compared with other PERKs. PERK5 members also contain an additional amino acid in the activation loop. The high sequence conservation within the PERK5 subfamily might indicate that these MAPKs are involved in some common signaling pathway and transmit related signals.

# 6
# PERK Groups: Towards a Functional Classification System of Plant MAPKs

Information on possible functions is available for members of the PERK1, 2, 3, and 4 subfamilies. The kinase activity of NtSIPK of the PERK1 subfamily is transiently activated by salicylic acid, wounding and by pathogens (Zhang and Klessig 1997 1998ab). Though NTF4 is highly similar to SIPK, it is most abundant in pollen and responds to rehydration during pollen germination (Wilson et al. 1997). The alfalfa member of the PERK1 subfamily, *MMK1* is transcriptionally upregulated in the G2-phase of the cell cycle, possibly indicating a function in cell cycle regulation (Jonak et al. 1993).

A characteristic feature for the members in the PERK2 subfamily is that a transcriptional upregulation of the gene accompanies the activation of the respective protein kinase. Some of these signals activate both PERK1 and PERK2 members, but there are also signals which appear to be subfamily-specific. WIPK, ATMPK3 and MMK4 all appear to be involved in mediating signaling of various stresses. The transcription and protein kinase activity of the MMK4 alfalfa MAP kinase was induced by cold, drought, wounding and mechanical stress (Bögre et al. 1996 1997; Jonak et al. 1996). The same stresses, including salt stress, induce the transcription of the ATMPK3 *Arabidopsis* MAP kinase gene (Mizoguchi et al. 1996). Some MAPKs are also activated by elicitors from different pathogens (Ligterink et al. 1997; Zhang and Klessig 1998b). With a reverse genetic approach it was shown that downregulation of WIPK mRNA levels in leaves interferes with wound-activation of MAPK and resulted in altered responses, abrogating the transcription of certain wound-induced genes and the production of jasmonic acid (Seo et al. 1995).

MMK2, a member of the PERK3 subfamily, is able to complement the yeast MPK1 mutation, involved in cell wall biosynthesis and actin organisation (Jonak et al. 1995). Consistent with a role in regulating actin organisation was the finding that a 39 kDa protein in cytoskeleton preparations is a substrate for MMK2 (Jonak et al. 1995).

The two known members of the PERK4 subfamily appear to function in mitosis. Both MMK3 and NTF6 are transiently activated as cells pass through ana- and telophase and were found to localise to the cell division plane in telophase cells (Calderini et al. 1998; Bögre et al. 1999). So far, no function for members of the PERK5 subfamily have been defined.

It should be noted that MAPKs almost certainly play a role in many other processes. Based on biochemical data, but without identification of the respective MAPK, a number of reports imply an involvement of MAPKs in the signal transduction of wounding (Usami et al. 1995; Stratman and Ryan 1997), pathogen defense (Suzuki and Shinshi 1995; Adam et al. 1997; Zhang et al. 1998), and several plant hormones. ABA is able to induce a MAPK-like activity in barley aleurone protoplasts (Knetsch et al. 1996). Besides inducing specific genes, ABA is known to inhibit giberellic acid (GA)-induced effects in aleurone cells. Whereas ABA stimulates MAPK activation in aleurone cells, GA may do the reverse, as indicated by the negative effect of GA on transcript accumulation of a MAPK gene in oat aleurone cells (Huttly and Phillips 1995).

Whereas auxin starvation arrests cell division in a tobacco cell suspension culture, readdition starts the cell cycle. During this process, a MAPKK and a protein kinase which has the properties of a MAPK are activated (Mizoguchi et al. 1994). A MAPK pathway may also be a negative regulator of auxin because expression of NPK1, a tobacco MAPKKK, in maize cells results in inhibition of auxin-induced gene expression (Kovtun et al. 1998).

Genetic analysis of the ethylene pathway in *Arabidopsis* indicates a possible involvement of a MAP kinase module. A number of mutants have been isolated that show a constitutive triple response (CTR) in the absence of ethylene. The *CTR1* gene was isolated and found to encode a protein with similarity to Raf1, a MAPKKK (Kieber et al. 1993).

# 7
# How Many Are Out There? Plant ESTs with Similarities to MAPKs

To estimate how many additional MAPKs might exist besides the 23 published full-length cDNAs, expressed sequence tag (EST) databases were searched with sequences from all five PERK subfamilies. In total, 45 ESTs were found that showed sequence similarity to MAPKs. Most of these ESTs are from *Arabidop-*

*sis* (24) and rice (9). All plant ESTs and partial cDNAs presently found in the databases are listed in Table 2 indicating length and position of the partial sequences relative to MMK1.

Out of the 24 *Arabidopsis* ESTs, 11 showed identity with the 7 published full-length clones. The other ESTs do not necessarily encode 13 other MAPKs in *Arabidopsis*, because these ESTs only encode N-terminal, central, or C-terminal MAPK regions that cannot be aligned with each other. The closest full-length MAPK homolog of each EST and three additional partial MAPK cDNA clones are also indicated in Table 2. Phylogenetic analysis of these sequences and the 21 full-length MAPKs revealed that the majority of the ESTs can be grouped into one of the five previously described PERK subfamilies (Fig. 4). At least one EST was found in each of the five subfamilies. Surprisingly, no *Arabidopsis* EST was found that could be grouped into the PERK4 subfamily. A reason for this could be that cDNAs for PERK4 members are not present or have a low abundance in the EST cDNA libraries. In contrast, many *Arabidopsis* ESTs were grouped in the PERK2 subfamily. In addition to the known full-length sequences, the EST analysis identified novel *Arabidopsis* MAPKs in the PERK2, 3 and 5 subfamilies.

Sixteen sequences (ten from Arabidopsis, two from cotton, one from rapeseed and rice, and two partial cDNA sequences from *S. lepidophylla*) could not be grouped to any of the five established PERK subfamilies. Instead, these sequences formed two new groups named EST1 and EST2 (Fig. 4). We are not

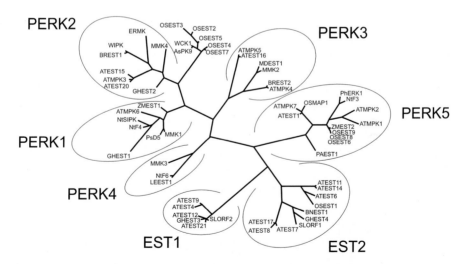

**Fig. 4.** Phylogenetic tree of all isolated plant MAPK ESTs, partial plant MAPK cDNAs and the 21 full-length plant MAPKs (Table 2). Subfamilies are marked as in Fig. 3. Two possible new MAPK groups, consisting only of EST and partial MAPK cDNAs are denoted as EST1 and EST2. The phylogenetic analysis was done with a heuristic search method as implemented in version 4.0.0d64 of PAUP* written by David L. Swofford

**Table 2.** Summary of all isolated plant MAPK EST and partial plant MAPK cDNAs

| Name | Accession No. | Species | Nucleotides (bp) | ORF (aa) | Closest homology | Position of orf (relative to MMK1) |
|---|---|---|---|---|---|---|
| AtEST1 | AA404767 | *Arabidopsis thaliana* | 476 | 73 | ATMPK7 | 316– |
| AtEST2 | AA605427 | *Arabidopsis thaliana* | 515 | 96 | ATMPK3 (ident.) | 296– |
| AtEST3 | AA721921 | *Arabidopsis thaliana* | 492 | 104 | ATMPK6 (ident.) | 1–96 |
| AtEST4 | AA713093 | *Arabidopsis thaliana* | 408 | 116 | SlORF2 | 35–152 |
| AtEST5 | H36168 | *Arabidopsis thaliana* | 466 | 150 | ATMPK3 (ident.) | 183–330 |
| AtEST6 | H37019 | *Arabidopsis thaliana* | 525 | 145 | SlORF1 | 355– |
| AtEST7 | N97150 | *Arabidopsis thaliana* | 535 | 105 | SlORF1 | 352– |
| AtEST8 | R90476 | *Arabidopsis thaliana* | 363 | 83 | FsMAPK (*Nectria haematococca*) | 314– |
| AtEST9 | N96333 | *Arabidopsis thaliana* | 510 | 133 | SlORF2 | 35–168 |
| AtEST10 | N65559 | *Arabidopsis thaliana* | 500 | 73 | ATMPK3 (ident.) | 20–90 |
| AtEST11 | T04165 | *Arabidopsis thaliana* | 429 | 119 | SlORF1 | 377– |
| AtEST12 | Z24542 | *Arabidopsis thaliana* | 323 | 104 | PMK1 (*Magnaporthe grisea*) | 196–296 |
| AtEST13 | R30629 | *Arabidopsis thaliana* | 466 | 102 | ATMPK2 (ident.) | 23–125 |
| AtEST14 | T21003 | *Arabidopsis thaliana* | 356 | 86 | SlORF1 | 334– |
| AtEST15 | R90032 | *Arabidopsis thaliana* | 400 | 64 | ATMPK3 | 20–82 |
| AtEST16 | T14181 | *Arabidopsis thaliana* | 359 | 115 | ATMPK5 (ident.) | 246–360 |
| AtEST17 | Z24497 | *Arabidopsis thaliana* | 306 | 59 | SlORF1 | 336– |
| AtEST18 | T21678 | *Arabidopsis thaliana* | 300 | 79 | ATMPK3 (ident.) | 132–208 |
| AtEST19 | T22107 | *Arabidopsis thaliana* | 371 | 134 | ATMPK5 (ident.) | 246–378 |
| AtEST20 | AI100610 | *Arabidopsis thaliana* | 375 | 54 | ATMPK3 | 331–386 |
| AtEST21 | T22988 | *Arabidopsis thaliana* | 501 | 126 | SpSPM1 (*S.pombe*) | 212–335 |
| AtEST22 | T41567 | *Arabidopsis thaliana* | 479 | 144 | ATMPK1 (ident.) | 23–167 |
| AtEST23 | T45144 | *Arabidopsis thaliana* | 482 | 164 | ATMPK3 (ident.) | 168–330 |
| AtEST24 | T46244 | *Arabidopsis thaliana* | 487 | 60 | ATMPK7 (ident.) | 23–83 |
| OsEST1 | AA750536 | *Oryza sativa* | 243 | 80 | SlORF | 379– |
| OsEST2 | AA753878 | *Oryza sativa* | 541 | 109 | WCK1 | 281–387 |
| OsEST3 | AA753910 | *Oryza sativa* | 498 | 95 | WCK1 | 295–387 |
| OsEST4 | C22363 | *Oryza sativa* | 462 | 153 | ASPK9 | 22–176 |

| | | | | | | |
|---|---|---|---|---|---|---|
| OsEST5 | C22364 | Oryza sativa | WCK1 | 463 | 84 | 304–387 |
| OsEST6 | C22686 | Oryza sativa | ATMPK1 | 463 | 40 | 350– |
| OsEST7 | C23546 | Oryza sativa | ASPK9 | 464 | 143 | 20–144 |
| OsEST8 | C23554 | Oryza sativa | NtF3 | 458 | 148 | 23–133 |
| OsEST9 | D41944 | Oryza sativa | ATMPK7 | 390 | 129 | 23–78 |
| ZmEST1 | AA979888 | Zea mays | PsD5 | 904 | 215 | 179– |
| ZmEST2 | T18821 | Zea mays | NtF3 | 272 | 90 | 58–148 |
| GhEST1 | AI054813 | Gossypium hirsutum | ATMPK6 | 596 | 140 | –3–116 |
| GhEST2 | AI055089 | Gossypium hirsutum | ERMK | 760 | 200 | 18–207 |
| GhEST3 | AI055345 | Gossypium hirsutum | FsMAPK (Nectria haematococca) | 621 | 154 | 174–322 |
| GhEST4 | AI055221 | Gossypium hirsutum | SlORF1 | 773 | 168 | 286– |
| BrEST1 | AT000688 | Brassica rapa ssp. pekinensis | NtF6 | 178 | 57 | 207–264 |
| BrEST2 | AT000946 | Brassica rapa ssp. pekinensis | ATMPK4 | 196 | 64 | 222–285 |
| BnEST1 | H07751 | Brassica napus | SlORF1 | 236 | 78 | 328– |
| MdEST1 | AT000385 | Malus domestica | MMK2 | 265 | 72 | 20–68 |
| PaEST1 | Z92691 | Picea abies | PhERK1 | 557 | 130 | 23–150 |
| LeEST1 | AA824827 | Lycopersicon esculentum | NtF6 | 603 | 69 | 319–387 |
| Other MAP kinase partial sequences | | | | | | |
| OsMAP1 | D21288 | Oryza sativa | NtF3 | 280 | 90 | 247–335 |
| SlORF1 | U96717 | Selaginella lepidophylla | NtF6 | 746 | 230 | 354– |
| SlORF2 | U96716 | Selaginella lepidophylla | PhERK1 | 709 | 196 | 38–231 |

certain whether the EST1 and EST2 groups really denote different subfamilies, because all the sequences in the EST1 group are derived from N-terminal or central MAPK parts, while all the members of the EST2 group originate from C-terminal MAPK sequences. No full-length MAPKs have been published which belong to these EST groups. Some members of the EST1 group (ATEST12, ATEST21, GHEST3, and SLORF2) contain the dual phosphorylation motif TDY in place of TEY that is present in all other PERKs. A TDY dual phosphorylation motif has so far only been found in MAPKs from protozoa. We also noted that members of the EST1 group (ATEST4, ATEST9, and SLORF2) do not contain the characteristic unique residue for other plant MAPKs (P39 in ATMPK3; Kültz 1998). Another feature for many members of the EST2 group is a long C-terminal extension, suggesting that these MAPKs will have a total length of between 434 and 560 amino acids, similar to members of the MAPK3 subfamily in animals, which are also substantially larger than other MAPKs (Kültz 1998).

A single signature sequence was found to be characteristic for all MAPKs and sufficient to distinguish MAPKs from other members of the protein kinase superfamily. This signature sequence is the 16 amino acid long sequence of the activaiton loop [L/I/V/M][T/S]XX[L/I/V/M]XT[R/K][W/Y]YRXPX[L/I/V/M][L/I/V/M], where the 2nd and 4th residues represent the phosphorylation sites (Kültz 1998). This signature does not hold for the putative MAPKs of the EST1 group. The sequence of members of this group is very conserved in this region and consists of the sequence WTDYVATRWYRXPELC. Both the first and the last residues do not fit with the general MAPK signature sequence.

Considering the complete sequence information on plant MAPKs, the PERK signature sequences proposed by Kültz (1998) have to be modified. The first proposed signature is ELMDTDLHXII[K/R]SXQ, located on the L7 loop. This signature is not valid for MMK3 which has a Q at position 12, and ATMPK1 which has a G instead of a Q at position 15. Modification of this sequence to ELMDTDLHXIIXS would give a signature sequence which would fit for every plant MAPK in the database, but is also present in some fungal MAPKs. The second proposed signature motif for plant MAPKs is HCX[F/ Y]F[L/I/V/M][F/Y]Q[L/I/V/M]LRGL[K/R]Y[L/I/V/M]HSAN located on the $\alpha_e$ helix. This signature is not valid for ATMPK1, which has a K instead of a G at position 12 of the signature. Modification of the signature sequence to HCX[F/Y]F[L/I/V/M][F/Y]Q[L/I/V/M]LRXL[K/R]Y[L/I/V/M]HSAN gives a signature sequence which is valid for all plant MAPKs in the database and is not present in any other protein. However, this signature sequence would probably fail to recognize MAPKs belonging to the EST1 and/or EST2 group. The only member of these groups that includes amino acid sequences in this region (SLORF2) has a different sequence. This again would suggest that the newly identified group(s) are not part of the general PERK subgroup of MAPKs.

Taken together, the grouping of plant MAPKs into subfamilies appears to provide a functional classification system. So far, a functional assignment has been based on the transcriptional upregulation of the respective MAPK gene or the activation of the kinase by defined stimuli. An understanding of the biological function of the MAPKs will require the identification of the targets and upstream kinases of the MAPKs and, last but not least, their context with other signaling pathways, appropriately termed neuronal networks, that make biological systems so wonderful.

# References

Adam AL, Pike S, Hoyos ME, Stone JM, Walker JC, Novacky A (1997) Rapid and transient activation of a myelin basic protein kinase in tobacco leaves treated with harpin from *Erwinia amylovora*. Plant Physiol 115:853–861

Bögre L, Ligterink W, Heberle-Bors E, Hirt H (1996) Mechanosensors in plants. Nature 383: 489–490

Bögre L, Ligterink W, Meskiene I, Barker PJ, Heberle-Bors E, Huskisson NS, Hirt H (1997) Wounding induces the rapid and transient activation of a specific MAP kinase pathway. Plant Cell 9:75–83

Bögre L, Calderini O, Binarova P, Mattauch M, Till S, Kiegerl S, Jonak C, Pollaschek C, Barker P, Huskisson NS, Hirt H, Heberle-Bors E (1999) A MAP kinase is activated late in plant mitosis and becomes localized to the plane of cell division. Plant Cell 11:101–114

Brunet A, Pouyssegur J (1996) Identification of MAP kinase domains by redirecting stress signals into growth factor responses. Science 272:1652–1655

Calderini O, Bögre L, Vicente O, Binarova P, Heberle-Bors E, Wilson C (1998) A cell cycle regulated MAP kinase with a possible role in cytokinesis in tobacco cells. J Cell Sci 111:3091–3100

Canagarajah BJ, Khokhlatchev A, Cobb MH, Goldsmith EJ (1997) Activation mechanism of the MAP kinase ERK2 by dual phosphorylation. Cell 90:859–869

Decroocq-Ferrant V, Decroocq S, Van Went J, Schmidt E, Kreis M (1995) A homolog of the MAP/ERK family of protein kinase genes is expressed in vegetative and in female reproductive organs of *Petunia hybrida*. Plant Mol Biol 27:339–350

Duerr B, Gawienowski M, Ropp T, Jacobs T (1993) MsERK1: a mitogen-activated protein kinase from a flowering plant. Plant Cell 5:87–96

Goldsmith EJ, Cobb MH (1994) Protein kinases. Curr Opin Struct Biol 4:833–840

Hanks SK, Hunter T (1995) The eukaryotic protein kinase superfamily: kinase (catalytic) domain structure and classification. FASEB J 9:576–596

Hanks SK, Quinn AM, Hunter T (1988) The protein kinase family: conserved features and deduced phylogeny of the catalytic domains. Science 241:42–52

Huttly A, Phillips AL (1995) Giberellin-regulated expression in oat aleurone cells of two kinases that show homology to MAP kinase and a ribosomal protein kinase. Plant Mol Biol 27: 1043–1052

Jiang Y, Li Z, Schwarz EM, Lin A, Guan K, Ulevitch RJ, Han J (1997) Structure-function studies of p38 mitogen-activated protein kinase. Loop 12 influences substrate specificity and autophosphorylation, but not upstream kinase selection. J Biol Chem 272:11096–11102

Jonak C, Páy A, Bögre L, Hirt H, Heberle-Bors E (1993) The plant homolog of MAP kinase is expressed in a cell cycle-dependent and organ specific manner. Plant J 3:611–617

Jonak C, Kiegerl S, Lloyd C, Chan J, Hirt H (1995) MMK2 a novel alfalfa MAP kinase, specifically complements the yeast MPK1 function. Mol Gen Genet 248:686–694

Jonak C, Kiegerl S, Ligterink W, Barker PJ, Huskisson NS, Hirt H (1996) Stress signalling in plants: a MAP kinase pathway is activated by cold and drought. Proc Natl Acad Sci USA 93: 11274–11279

Kieber JJ, Rothenberg M, Roman G, Feldmann KA, Ecker JR (1993) CTR1, a negative regulator of the ethylene response pathway in *Arabidopsis*, encodes a member of the Raf family of protein kinases. Cell 72:427–441

Knetsch MLW, Wang M, Snaar-Jagalska BE, Heimovaara-Dijkstra S (1996) Abscisic acid induces mitogen-activated protein kinase activation in barley aleurone protoplasts. Plant Cell 8: 1061–1067

Kovtun Y, Chiu W-L, Zeng W, Sheen J (1998) Suppression of auxin signal transduction by a MAPK cascade in higher plants. Nature 395:716–720

Kültz D (1998) Phylogenetic and functional classification of mitogen- and stress-activated protein kinases. J Mol Evol 46:571–588

Ligterink W, Kroj T, zur Nieden U, Hirt H, Scheel D (1997) Receptor-mediated activation of a MAP kinase in pathogen defense of plants. Science 276:2054–2057

Madhani HD, Fink GR (1998) The riddle of MAP kinase signaling specificity. Trends Genet 14: 151–155

Marshall CJ (1994) MAP kinase kinase kinase, MAP kinase kinase, and MAP kinase. Curr Op in Gen Dev 4:82–89

Mizoguchi T, Hayashida N, Yamaguchi-Shinozaki K, Kamada H, Shinozaki K (1993) ATMPKs: a gene family of plant MAP kinases in *Arabidopsis thaliana*. FEBS Lett 336:440–444

Mizoguchi T, Gotoh Y, Nishida E, Yamaguchi-Shinozaki K, Hayashida N, Iwasaki T, Kamada H, Shinozaki K (1994) Characterization of two cDNAs that encode MAP kinase homologues in *Arabidopsis thaliana* and analysis of the possible role of auxin in activating such kinase activities in cultured cells. Plant J 5:111–122

Mizoguchi T, Irie K, Hirayama T, Hayashida N, Yamaguchi-Shinozaki K, Matsumoto K, Shinozaki K (1996) A gene encoding a mitogen-activated protein kinase kinase kinase is induced simultaneously with genes for a mitogen-activated protein kinase and S6 ribosomal protein kinase by touch, cold, and water stress in *Arabidopsis thaliana*. Proc Natl Acad Sci USA 93: 765–769

Mizoguchi T, Ichimura K, Shinozaki K (1997) Environmental stress response in plants: the role of mitogen-activated protein kinases. Trends Biotechnol 15:15–19

Mizoguchi T, Ichimura K, Irie K, Morris P, Giraudat J, Matsumoto K, Shinozaki K (1998) Identification of a possible MAP kinase cascade in *Arabidopsis thaliana* based on pairwise yeast two-hybrid analysis and functional complementation tests of yeast mutants. FEBS Lett 437: 56–60

Seo S, Okamoto M, Seto H, Ishizuka K, Sano H, Ohashi Y (1995) Tobacco MAP kinase: a possible mediator in wound signal transduction pathways. Science 270:1988–1992

Stafstrom JP, Altschuler M, Anderson DH (1993) Molecular cloning and expression of a MAP kinase homologue from pea. Plant Mol Biol 22:83–90

Stratmann JW, Ryan CA (1997) Myelin basic protein kinase activity in tomato leaves is induced systemically by wounding and increases in response to systemin and oligosaccharide elicitors. Proc Natl Acad Sci USA 94:11085–11089

Suzuki K, Shinshi H (1995) Transient activation and tyrosine phosphorylation of a protein kinase in tobacco cells treated with fungal elicitor. Plant Cell 7:639–647

Takezawa D (1998) A23187-induced mRNA expression of WCK-1, a gene encoding mitogen-activated protein kinase in plants. (unpublished)

Taylor SS, Radzio-Andzelm E (1994) Three protein kinase structures define a common motif. Structure 5:345–355

Usami S, Banno H, Ito Y, Nishihama R, Machida Y (1995) Cutting activates a 46-kDa protein kinase in plants. Proc Natl Acad Sci USA 92:8660–8664

Wang Z, Harkins PC, Ulevitch RJ, Han J, Cobb MH, Goldsmith EJ (1997) The structure of mitogen-activated protein kinase p38 at 2.1-Å resolution. Proc Natl Acad Sci USA 94:2327–2332

Wilson C, Eller N, Gartner A, Vicente O, Heberle-Bors E (1993) Isolation and characterization of a tobacco cDNA clone encoding a putative MAP kinase. Plant Mol Biol 23:543–551

Wilson C, Anglmayer R, Vicente O, Heberle-Bors E (1995) Molecular cloning, functional expression in *Escherichia coli*, and characterization of multiple mitogen-activated-protein kinases from tobacco. Eur J Biochem 233:249–257

Wilson C, Voronin V, Touraev A, Vicente O, Heberle-Bors E (1997) A developmentally regulated MAP kinase activated by hydration in tobacco pollen. Plant Cell 9:2093–2100

Zhang F, Strand A, Robbins D, Cobb MH, Goldsmith EJ (1994) Atomic structure of the MAP kinase ERK2 at 2.3 Å resolution. Nature 367:704–711

Zhang J, Zhang F, Ebert D, Cobb MH, Goldsmith EJ (1995) Activity of the MAP kinase ERK2 is controlled by a flexible surface loop. Structure 3:299–307

Zhang S, Klessig DF (1997) Salicylic acid activates a 48 kDa MAP kinase in tobacco. Plant Cell 9:809–824

Zhang S, Klessig DF (1998a) The tobacco wounding-activated mitogen-activated protein kinase is encoded by SIPK. Proc Natl Acad Sci USA 95:7225–7230

Zhang S, Klessig DF (1998b) Resistance gene N-mediated de novo synthesis and activation of a tobacco mitogen-activated protein kinase by tobacco mosaic virus infection. Proc Natl Acad Sci USA 95:7433–7438

Zhang S, Du H, Klessig DF (1998) Activation of the tobacco SIP kinase by both a cell wall-derived carbohydrate elicitor and purified proteinaceous elicitins from *Phythophthora* spp. Plant Cell 10:435–449

# MAP Kinase Cascades in *Arabidopsis*: Their Roles in Stress and Hormone Responses

Tsuyoshi Mizoguchi[1], Kazuya Ichimura[1,2], Riichiro Yoshida[1], and Kazuo Shinozaki[1]*

**Summary:** Mitogen-activated protein kinase (MAPK) cascades have essential roles in diverse intracellular signaling processes in plants, animals and yeasts. In plants, MAPK and MAPK-like kinase activities are transiently activated in response to environmental stresses and plant hormones. In addition, transcription of genes encoding protein kinases involved in MAPK cascades is upregulated by environmental stresses. A possible MAPK cascade of *Arabidopsis* was identified based on both the yeast two-hybrid analysis and functional complementation analysis of yeast mutants. This MAPK cascade may have important roles in stress signal transduction pathways in *Arabidopsis*.

# 1
# Introduction

Mitogen-activated protein kinases (MAPKs) were initially identified as serine/threonine kinases that were activated by various growth factors, differentiation and mitotic signals in animals (for reviews, see Nishida and Gotoh 1993; Cano and Mahadevan 1995; Herskowitz 1995). MAPKs have key roles in integrating multiple intracellular signals transmitted by various second messengers. MAPK cascades are composed of three protein kinases, MAPKs, MAPK kinases (MAPKKs), and MAPKK kinases (MAPKKKs, Fig. 1). MAPKs are activated when both tyrosine and threonine residues in the TXY motif are phosphorylated by MAPKKs. MAPKKs are activated when serine and serine/threonine residues in the SXXXS/T motif are phosphorylated by MAPKKKs. MAPK cascades have been shown to function in various signal transduction pathways

* *Corresponding author:* Dr. Kazuo Shinozaki
  Laboratory of Plant Molecular Biology, The Institute of Physical and Chemical Research
  (RIKEN), Tsukuba Life Science Center, 3-1-1, Koyadai, Tsukuba, Ibaraki 305-0074, Japan
[1] Laboratory of Plant Molecular Biology, The Institute of Physical and Chemical Research
  (RIKEN), Tsukuba Life Science Center, 3-1-1, Koyadai, Tsukuba, Ibaraki 305-0074, Japan
[2] Institute of Biological Sciences, The University of Tsukuba, Tennodai, Tsukuba,
  Ibaraki 305-0006, Japan

Results and Problems in Cell Differentiation, Vol. 27
Hirt (Ed.): MAP Kinases in Plant Signal Transduction
© Springer-Verlag Berlin Heidelberg 2000

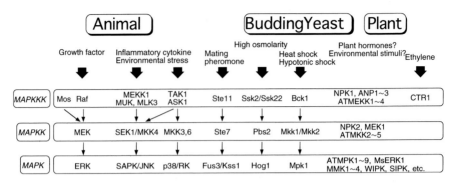

**Fig. 1.** A hypothetical model for MAPK cascades in various signal transduction pathways in animals, budding yeasts and plants. Numerous MAPKKK/MAPKK/MAPK homologs have been identified in plants. Other details are given in the text

in animals and yeasts (for reviews, see Nishida and Gotoh 1993; Jonak et al. 1994; Cano and Mahadevan 1995; Herskowitz 1995; Nishihama et al. 1995; Hirt 1997; Mizoguchi et al. 1997). In plants, many cDNAs for MAPKs, MAPKKs, and MAPKKKs have been isolated from various species (for reviews, see Jonak et al. 1994; Nishihama et al. 1995; Hirt 1997; Mizoguchi et al. 1997). A larger number of MAPKs have been identified in plants than in animals and yeasts, which suggests that MAPK cascades have a wider variety of roles in plants. Their functions, however, have not been elucidated. Recently, it has been shown that mRNA levels of the components of MAPK cascades increase in response to various environmental stressors in plants. Activation of MAPKs activity in response to environmental stresses have been analyzed using specific antibodies. These results suggest that some of the plant MAPK cascades play important roles in environmental stress responses as well as in plant hormone signal transduction.

## 2
## MAPK Cascades Involved in Stress Responses in Animals and Budding Yeast

In animal cells, at least three different MAPK cascades function in the different cellular signaling pathways (Fig. 1). A cascade involving Raf/Mos, MEK1/ MEK2 and ERK1 operates in growth-factor-dependent cell proliferation. The response to stresses such as UV irradiation, translational inhibitors, heat shock and tumor necrosis factor involves a MAPK cascade comprising MEKK, SEK/MKK4 and SAPK/JNK, whereas inflammatory cytokines (tumor necrosis factor and interleukin-1) and environmental stresses (osmotic shock and UV irradiation) involve MAPK cascades, MKK3/MKK6/MKK4 and p38 (Nishida and Gotoh 1993; Cano and Mahadevan 1995; Herskowitz 1995; Raingeaud et al. 1996). In budding yeast, at least five different MAPK signaling pathways have been described (Cano and Mahadevan 1995; Herskowitz 1995; Madhani and Fink 1998). They are the mating-pheromone response pathway (Ste11, Ste7, and Fus3/Kss1), the protein kinase C (PKC)-dependent signaling pathway involved in maintenance of cell wall integrity (Bck1, Mkk1/Mkk2, and Mpk1), the osmoregulatory pathway (Ssk2/Ssk22, Pbs2, and Hog1), the pseudohyphal differentiation pathway (Ste11 and Ste7), and the spore wall assembly pathway (Smk1).

# 3
# *Arabidopsis* Homologs of Protein Kinases in MAPK Cascades

A number of genes for MAPKs have been reported in higher plants (for reviews, see Jonak et al. 1994; Nishihama et al. 1995; Hirt 1997; Mizoguchi et al. 1997). We have demonstrated that MAPKs constitute a gene family with at least nine members (ATMPK1–9) in *Arabidopsis thaliana.* (Mizoguchi et al. 1993, 1994, 1997). A phylogenetic tree (Fig. 2) indicates the evolutional relationship

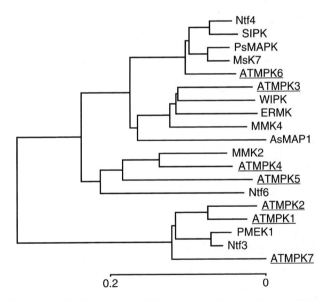

**Fig. 2.** Evolutional relationship among MAPKs in plants. These plant MAPKs can be classified into four subgroups. The evolutional distance between one protein and another is indicated at the *bottom.* The accession numbers in the DDBJ, EMBL and GenBank databases of the plant MAPKs in this figure are as follows: Ntf4 (X83880), SIPK (U94192), PsMAPK (X70703), MsK7 (L07042), ATMPK6 (D21842), ATMPK3 (D21839), WIPK (D61377), ERMK (Y12875), MMK4 (X82270), AsMAP (X79993), MMK2 (X82268), ATMPK4 (D21840), ATMPK5 (D21841), Ntf6 (X83879), ATMPK2 (D14714), ATMPK1 (D14713), PMEK1 (X83440), Ntf3 (X69971) and ATMPK7 (D21843)

among MAP kinases in plants. Plant MAPKs, including seven *Arabidopsis* MAPKs (ATMPK1–7) can be classified into the following four subgroups based on phylogenetic analysis of their amino acid sequences: subgroup 1, ATMPK1, ATMPK2, ATMPK7, PMEK1 and Ntf3; subgroup 2, ATMPK3, WIPK, ERMK, MMK4 and AsMAP1; subgroup 3, ATMPK6, Ntf4, SIPK, PsMAPK and MsK7; subgroup 4, ATMPK4, ATMPK5, MMK2 and Ntf6. ATMPK8 and ATMPK9 are classified into the fifth new subgroup. Instead of the motif TEY on animal ERKs, ATMPK8 and ATMPK9 contain the sequence TDY (Mizoguchi et al. 1997). We searched for additional MAPK genes in the *Arabidopsis* genome using PCR with specific primers for the MAPK family, plaque hybridization screening under low-stringency conditions, and sequence information of *Arabidopsis* ESTs in databases. *Arabidopsis* may have at least 30 MAPK genes (R. Yoshida, T. Mizoguchi, K. Shinozaki, unpubl. data).

In *Arabidpsis*, five protein kinases (MEK1, ATMKK2–5) related to MAPKKs have been reported (Morris et al. 1997; Ichimura et al. 1998a). Plant MAPKK homologs can be classified into the following three subgroups based on phylogenetic analysis of their sequences: subgroup 1, NPK2 (a tobacco MAPKK homolog; Shibata et al. 1995) and ATMKK3; subgroup 2, MEK1 and ATMKK2; subgroup 3, ATMKK4 and ATMKK5 (Ichimura et al. 1998a).

Three types of protein kinases related to MAPKKKs have been isolated from *Arabidopsis*. The CTR1 gene, encoding a Raf homolog that functions as an activator of MAPKK in animal cells, has a key role in the ethylene signal transduction pathway (Kieber et al. 1993). We isolated *Arabidopsis* cDNAs (ATMEKK1–4) encoding protein kinases which are structurally related to MAPKKKs (such as Byr2, Ste11, Bck1, MEKK, and NPK1; Mizoguchi et al. 1996). ATMEKK1 can complement the yeast *ste11* mutant in response to the mating pheromone (Mizoguchi et al. 1996). Nishihama *et al.* have isolated three cDNAs (ANP1–3) from *Arabidopsis* which encode homologs of NPK1 (Nishihama et al. 1997). NPK1, which is structurally related to MAPKKKs, was isolated from tobacco and shown to complement the growth defect of the yeast *bck1Δ* mutant (Banno et al. 1993).

## 4
## Transcriptional Control of the Putative Components of *Arabidopsis* MAPK Cascades in Response to Environmental Stresses

We demonstrated that the transcription levels of three protein kinases, ATMEKK1 (a MAPKKK), ATMPK3 (a MAPK), and ATPK19 (a Ribosomal S6 kinase or RSK homolog), increased markedly and simultaneously when plants were treated with touch stimuli, low temperature, and salinity stress (Mizoguchi et al. 1996). These results suggest that some of the MAPK cascades in plants might function in transducing signals in the presence of environmental stress, and that MAPK cascades are regulated at a transcriptional as well as a post-translational level in plants (Mizoguchi et al. 1996). Seo et al. isolated a tobacco cDNA clone for a wound-inducible gene encoding a MAPK homolog in tobacco (WIPK; Seo et al. 1995). The transcript for WIPK increased rapidly after mechanical wounding. The transcript level as well as the kinase activity of MMK4 increased after cold, drought and wounding treatments of alfalfa plants (Jonak et al. 1996; Bögre et al. 1996b). Four MAPKs from different plant species, ATMPK3, WIPK, MMK4 and ERMK (subgroup 2 of plant MAPK family) have a higher sequence identity to one another than to other MAPKs, and the mRNA levels of these MAPKs increased after stress treatments, suggesting that these MAPKs may play similar roles in plants under a variety of stress conditions (Seo et al. 1995; Jonak et al. 1996; Mizoguchi et al. 1996; Bögre et al. 1997; Ligterink et al. 1997). In higher plants, mRNAs of various genes involved in signal transduction pathways, such as protein kinases, a phospholipase C, a histidine kinase, monomeric small GTP binding proteins, calmodulins, and transcription factors, accumulate in response to environmental stimuli or stresses (Shinozaki and Yamaguchi-Shinozaki 1997). Under environmental stress conditions, the elevated levels of the mRNAs of these putative signal transducers may increase their protein level, which is likely to amplify the signal transduction efficiency of the cascade.

# 5
# Activation of MAPK-Like Activity by Environmental Stresses in *Arabidopsis thaliana*

In animal cells, cellular stresses (osmotic shock, heat shock, ATP depletion, UV- and gamma radiation) and inflammatory cytokines (TNF-a and IL-1) induce activation of SAPK/JNK and p38 MAPKs (Cano and Mahadevan 1995; Herskowitz 1995; Raingeaud et al. 1996). In budding yeast, Hog1 MAPK is activated by hyperosmotic stress and Mpk1 MAPK is activated by heat shock and hypotonic shock (Brewster et al. 1993; Kamada et al. 1995). In plants, Usami et al. reported that cutting mature tobacco leaves activated a 46-kDa MAPK-like protein kinase and that activity was stimulated not only tobacco, but also in *Arabidopsis* and various other plants (Usami et al. 1995). Biochemical analysis using a specific antibody for MMK4, an Alfalfa MAPK homologue, showed that MMK4 was activated by wounding, mechanical stress, cold and drought (Jonak et al. 1996; Bögre et al. 1996a, b). The Amino acid sequence of MMK4 is closely related to an *Arabidopsis* ATMPK3, tobacco WIPK and parsley ERMK (Mizoguchi et al. 1993; Seo et al. 1995; Jonak et al. 1996; Ligterink et al. 1997). Transcription of ATMPK3 is rapidly induced by touch, cold and water stress (Mizoguchi et al. 1996). Gene expression of MMK4 is also up-regulated by wounding, cold and drought (Bögre et al. 1996b; Jonak et al. 1996). Structures and expression patterns of ATMPK3 and MMK4 are very similar, which suggests

ATMPK3 may be activated by wounding, touch, cold and water stress in *Arabidopsis*.

On the other hand, we detected wounding-, touch-, cold- and dehumidification-stimulated rapid and transient activation of several protein kinases (47–40 kDa in size) which phosphorylate myelin basic protein (MBP) in gel phosphorylation assays in *Arabidopsis* (K. Ichimura, T. Mizoguchi, K. Shinozaki, unpubl. data). These protein kinases seem to be MAPKs based on positive immuno-precipitation with antibodies for phosphotyrosine and the TEY motif. Recently, we have demonstrated that one of the stress-activated MAPKs cross reacts with a specific antibody for ATMPK4, which suggests ATMPK4 may be involved in stress responses.

Tobacco SIPK is initially identified as salicylic acid activated protein kinase and purified from tobacco cultured cells (Zhang and Klessig 1997). Biochemical analysis using a specific antibody showed that SIPK was also activated by elicitors and wounding (Zhang and Klessig 1998a, b). In Fig. 2, SIPK can be classified into subgroup 3, which is different from the subgroup containing MMK4, WIPK and ATMPK3. ATMPK6 is classified into the same subgroup as SIPK. Members of MAPKs in subgroup 3 reveal a slightly larger size (around 48kDa) than other subfamilies (around 43kDa), except the subgroup 5 of ATMPK8 and ATMPK9. A putative molecular size of ATMPK6 is 47kDa. ATMPK6 as well as ATMPK3 and ATMPK4 may also be involved in stress responses, like SIPK.

# 6
# Analysis of Functional Interactions of the Components of MAPK Cascades in *Arabidopsis* Using the Yeast System

A number of genes for MAPKs and several genes for MAPKKs and MAPKKKs have been reported in higher plants (for reviews, see Jonak et al. 1994; Nishihama et al. 1995; Hirt 1997; Mizoguchi et al. 1997). However, there have been no reports on the demonstration of direct interactions between MAPK and MAPKK, or between MAPKK and MAPKKK from plants, and direct phosphorylation and activation of either MAPK by MAPKK, or MAPKK by MAPKKK from plants. Recently, protein-protein interactions between ATMEKK1 (a MAPKKK) and MEK1/ATMKK2 (MAPKKs), between MEK1/ ATMKK2 (-MAPKKs) and ATMPK4 (a MAPK) and between ATMPK4 (a MAPK) and AT-MEKK1 (a MAPKKK) were detected by the yeast two-hybrid system (Mizoguchi et al. 1997, 1998; Ichimura et al. 1998b). Co-expression of ATMEKK1 (a MAPKKK) increased the ability of two closely related MAPKKs, MEK1 and ATMKK2, to complement the growth defect of the *pbs2* mutant (Mizoguchi et al. 1997, 1998; Ichimura et al. 1998b). Co-expression of ATMPK4 (a MAPK) and MEK1 (a MAPKK) could complement the growth defect of the *mpk1* and the *bck1* mutants, although co-expression with ATMKK2 could not. By contrast, other combinations could not suppress these mutaions. These results suggest that ATMPK4 (a MAPK), MEK1/ATMKK2 (MAPKKs) and ATMEKK1 (a MAPKKK) may constitute a MAP kinase cascade and that two closely related MAPKKs, MEK1 and ATMKK2, may have similar roles in *Arabidopsis.*

# 7
# Putative Upstream Factors of MAPK Cascades in *Arabidopsis*

In bacteria, the two-component system functions in sensing and response to osmotic stress (Wurgler-Murphy and Saito 1997). The system is composed of two types of poteins, a sensory histidine kinase (EnvZ) and a response regulator (OmpR). In yeast, exposure to high osmolarity activate MAPK cascades, which includes Ssk2/Ssk22 (MAPKKKs), Pbs2 (a MAPKK) and Hog1 (a MAPK), and then activates several genes involved in the biosynthesis of glycerol, which is an important osmoprotectant (Fig. 3; Wurgler-Murphy and Saito 1997). Three gene products (Sln1, Ypd1 and Ssk1) which act in an early phase of the hyperosmolarity-stress response encode signaling molecules that constitute a prokaryote-type two-component regulatory system (Posas et al. 1996). Sln1 is thought to act as a sensor protein, phosphorylating Ypd1 and Ssk1 response regulator proteins under conditions of high osmolarity. The three protein factors perform a four step phospho-relay (His-Asp-His-Asp). At high osmolarity, unphosphorylated Ssk1 activates Ssk2 or Ssk22 (MAPKKKs), which results in the activation of Pbs2 (a MAPKK) by Ser-Thr phosphorylation (Maeda et al. 1995). Then phosphorylated Pbs2 activates Hog1 (a MAPK) by Thr-Tyr phosphorylation (Fig. 3).

Recently, we have isolated cDNAs encoding a two-component histidine kinase (ATHK1), phospho-relay intermediate Ypd1 homologs which contain a Hpt domain (ATHPs) and response regulator homologs (ATRRs) from *Arabidopsis* (Miyata et al. 1998; Urao et al. 1998, 1999 submitted). Overexpression of ATHK1 suppressed the growth defect of the yeast *sln1* mutant with a deficient osmosensor histidine kinase. The ATHK1 transcript accumulated in conditions of high salinity and low temperature. The existence of a two-component system and MAP kinase cascades which may be involved in stress signaling in *Arabidopsis* suggests that plants have an osmosensing and signaling system similar to that of yeast (Fig. 3). In *Arabidopsis*, a two-component histidine kinase, ETR1, is a receptor in ethylene signal transduction, and another two-component histidine kinase, CKI1, is involved in cytokinin signaling (Chang et al. 1993; Kakimoto 1996). CTR1 which has high homology to Raf (a mammalian MAPKKK) has recently been shown to interact directly with ETR1 (Clark et al. 1998). Two-component histidine kinases may function as sensors or receptors in various signal-transduction pathways and as upstream factors of MAP kinase cascades in plants, as in yeast. Analyses of transgenic plants that overexpress or repress genes for factors of MAP kinase cascades and two-com-

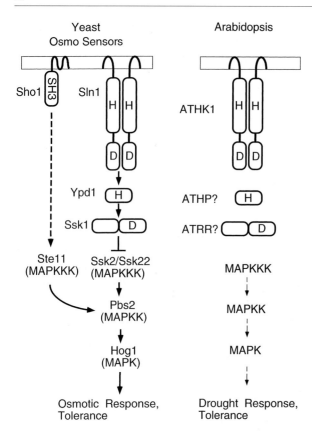

Fig. 3. Possible roles of the two-component system and MAP kinase cascades in plants compared with the yeast osmosensing system. In yeast, Sln1 is thought to act as a sensor protein, phosphorylating Ypd1 (a phospho-relay intermediate) and Ssk1 (a response regulator), under conditions of high osmolarity. A MAP kinase cascade, Ssk2/Ssk22 (MAPKKKs), Pbs2 (a MAPKK) and Hog1 (a MAPK), functions downstream of the two-component system in osmotic-stress response. ATHK1, ATHP and ATRR from *Arabidopsis* are proteins related to a sensor histidine kinase, a phospho-relay intermediate and a response regulator, respectively. A similar two-component system and a MAP kinase cascade may be involved in drought-stress signal transduction. See text for details

ponent systems, will help to elucidate the molecular mechanisms of the stress response and adaptation to environmental changes in higher plants. Isolation of insertion mutants of these genes which contain T-DNA or transposons will be useful for analyzing the function of MAPKs and two-component systems in plants.

## References

Banno H, Hirano K, Nakamura T, Irie K, Nomoto S, Matsumoto K, Machida Y (1993) NPK1, a tobacco gene that encodes a protein with a domain homologous to yeast BCK1, STE11 and Byr2 protein kinases. Mol Cell Biol 13:4745–4752

Bögre L, Ligterink W, Heberle-Bors E, Hirt H (1996a) Mechanosensors in plants. Nature 383:489–490

Bögre L, Ligterink W, Meskience I, Barker PJ, Heberle-Bors E, Huskisson NS, Hirt H (1996b) Wounding induces the rapid and transient activation of a specific MAP kinase pathway. Plant Cell 9:75–83

Brewster JL, de Valoir T, Dwyer ND, Winter E, Gustin MC (1993) An osmosensing signal transduction pathway in yeast. Science 259:1760–1763

Cano E, Mahadevan LC (1995) Parallel signal processing among mammalian MAPKs.TIBS 20: 117–122

Chang C, Kwok SF, Bleecker AB, Meyerowitz EM (1993) *Arabidopsis* ethylene-response gene ETR1: similarity of product to two-component regulators. Science 262:539–544

Clark KL, Larsen PB, Wang X, Chang C (1998) Association of the *Arabidopsis* CTR1 Raf-like kinase with the ETR1 and ERS ethylene receptors. Proc Natl Acad Sci USA 95:5401–5406

Herskowitz I (1995) MAP kinase pathways in yeast: for mating and more. Cell 80:187–197

Hirt H (1997) Multiple roles of MAP kinases in plant signal transduction. Plant Sci 2:11–15

Ichimura K, Mizoguchi T, Hayashida N, Seki M, Shinozaki K (1998a) Molecular cloning and characterization of three cDNAs encoding putative mitogen-activated protein kinase kinases (MAPKKs) in *Arabidopsis thaliana*. DNA Res 5:341–348

Ichimura K, Mizoguchi T, Irie K, Morris P, Giraudat J, Matsumoto K, Shinozaki K (1998b) Isolation of ATMEKK1 (a MAP kinase kinase kinase)-interacting proteins and analysis of a MAP kinase cascade in *Arabidopsis*. Biochem Biophys Res Comm 253:532–543

Jonak C, Heberle-Bors E, Hirt H (1994) MAP kinases: universal multi-purpose signaling tools. Plant Mol Biol 24:407–416

Jonak C, Kiegel S, Ligterink W, Barker PJ, Huskisson NS, Hirt H (1996) Stress signaling in plants: a mitogen-activated protein kinase pathway is activated by cold and drought. Proc Natl Acad Sci USA 93:11274–11279

Kakimoto T (1996) CKI1, a histidine kinase homologue implicated in cytokinin signal transduction. Science 274:982–985

Kamada Y, Jung US, Piotrowski J, Levin DE (1995) The protein kinase C-activated MAP kinase pathway of *Saccharomyces cerevisiae* mediates a novel aspect of the heat shock response. Genes Dev 13:1559–1571

Kieber JJ, Rothenberg M, Roman G, Feldmann KA, Ecker JR (1993) *CTR1*: a negative regulator of the ethylene response pathway in *Arabidopsis*, encodes a member of the Raf family of protein kinases. Cell 72:427–441

Ligterink W, Kroj T, Nieden UZ, Hirt H, Scheel D (1997) Receptor-mediated activation of a MAP kinase in pathogen defense of plants. Science 276:2054–2057

Madhani HD, Fink GR (1998) The riddle of MAP kinase signaling specificity. Trends Genet 14: 151–155

Maeda T, Takekawa M, Saito H (1995) Activation of yeast PBS2 MAPKK by MAPKKKs or by binding of an SH3-containing osmosensor. Science 269:554–558

Miyata S, Urao T, Yamaguchi-Shinozaki K, Shinozaki K (1998) Characterization of genes for two-component phosphorelay mediators with a single Hpt domain in *Arabidopsis thaliana*. FEBS Lett 437:11–14

Mizoguchi T, Hayashida N, Yamaguchi-Shinozaki K, Kamada H, Shinozaki K (1993) ATMPKs: a gene family of plant MAP kinases in *Arabidopsis thaliana*. FEBS Lett 336:440–444

Mizoguchi T, Gotoh Y, Nishida E, Yamaguchi-Shinozaki K, Hayashida N, Iwasaki T, Kamada H, Shinozaki K (1994) Characterization of two cDNAs that encode MAP kinase homologues in *Arabidopsis thaliana* and analysis of the possible role of auxin in activating such kinase activities in cultured cells. Plant J 5:111–122

Mizoguchi T, Irie K, Hirayama T, Hayashida N, Yamaguchi-Shinozaki K, Matsumoto K, Shinozaki K (1996) A gene encoding a mitogen-activated protein kinase kinase kinase is induced simultaneously with genes for a mitogen-activated protein kinase and an S6 ribosomal protein kinase by touch, cold, and water stress in *Arabidopsis thaliana*. Proc Natl Acad Sci USA 93: 765–769

Mizoguchi T, Ichimura K, Shinozaki K (1997) Environmental stress response in plants: the role of mitogen-activated protein kinases. Trends Biotechnol 15:15–19

Mizoguchi T, Ichimura K, Irie K, Morris P, Giraudat J, Matsumoto K, Shinozaki K (1998) Identification of a possible MAP kinase cascade in *Arabidopsis thaliana* based on pairwise yeast two-hybrid analysis and functional complementation tests of yeast mutants. FEBS Lett 437:56–60

Morris PC, Guerrier D, Leung J, Giraudat J (1997) Cloning and characterization of *MEK1*, an *Arabidopsis* gene encoding a homologue of MAP kinase kinase. Plant Mol Biol 35:1057–1064

Nishida E, Gotoh Y (1993) The MAP kinase cascade is essential for diverse signal transduction pathways. TIBS 18:128–131

Nishihama R, Banno H, Shibata W, Hirano K, Nakashima M, Usami S, Machida Y (1995) Plant homologues of components of MAPK (mitogen-activated protein kinase) signal pathways in yeast and animal cells. Plant Cell Physiol 36:749–757

Nishihama R, Banno H, Kawahara E, Irie K, Machida Y (1997) Possible involvement of differential splicing in regulation of the activity of *Arabidopsis* ANP1 that is related to mitogen-activated protein kinase kinase kinases (MAPKKKs). Plant J 12:39–48

Posas F, Wurgler-Murphy SM, Maeda T, Witten EA, Thai TC, Saito H (1996) Yeast HOG1 MAP kinase cascade is regulated by a multistep phosphorelay mechanism in the SLN1-YPD1-SSK1 "two-component" osmosensor. Cell 86:865–875

Posas F, Saito H (1997) Osmotic activation of the HOG MAPK pathway via Ste11p MAPKKK: scaffold role of Pbs2p MAPKK. Science 276:1702–1705

Raingeaud J, Whitmarsh AJ, Barret T, Dérijard B, Davis RJ (1996) MKK3- and MKK6-regulated gene expression is mediated by the p38 mitogen-activated protein kinase signal transduction pathway. Mol Cell Biol 16:1247–1255

Seo S, Okamoto M, Seto H, Ishizuka K, Sano H, Ohashi Y (1995) Tobacco MAP kinase: a possible mediator in wound signal transduction pathways. Science 270:1988–1992

Shibata W, Banno H, Ito Y, Hirano K, Irie K, Usami S, Machida C, Machida Y (1995) A tobacco protein kinase, NPK2, has a domain homologous to a domain found in activators of mitogen-activated protein kinases (MAPKKs). Mol Gen Genet 246:401–410

Shinozaki K, Yamaguchi-Shinozaki K (1997) Gene expression and signal transduction in water-stress response. Plant Physiol 115:327–334

Urao T, Yakubov B, Yamaguchi-Shinozaki K, Shinozaki K (1998) Stress-responsive expression of genes for two-component response regulator-like proteins in *Arabidopsis thaliana*. FEBS Lett 427:175–178

Urao T, Yakubov B, Satoh R, Yamaguchi-Shinozaki K, Seki M, Hirayama T, Shinozaki K (1999) A transmembrane hybrid-type histidine kinase in *Arabidopsis* functions as an osmosensor. Plant Cell, in press

Usami S, Banno H, Ito Y, Nishihama R, Machida Y (1995) Cutting activates a 46kDa protein kinase in plants. Proc Natl Acad Sci USA 92:8660–8664

Wurgler-Murphy SM, Saito H (1997) Two-component signal transducers and MAPK cascades. TIBS 22:172–176

Zhang S, Klessig DF (1997) Salicylic acid activates a 48-kD MAP kinase in tobacco. Plant Cell 9:809–824

Zhang S, Klessig DF (1998a) Activation of the tobacco SIP kinase by both a cell wall-derived carbohydrate elicitor and purified proteinaceous elicitins from *Phytophthora* spp. Plant Cell 9:809–824

Zhang S, Klessig DF (1998b) The tobacco wounding-activated mitogen-activated protein kinase is encoded by *SIPK*. Proc Natl Acad Sci USA 95:7225–7230

# MAP Kinases in Pollen

Cathal Wilson[1] and Erwin Heberle-Bors[1][*]

**Summary:** The male gametophyte of flowering plants has a highly regulated developmental programme to ensure efficient fertilization of the ovule and the faithful transmission of genetic material to the offspring. Cell cycle control mechanisms dictate the formation of the vegatative and generative (sperm) cells, while an increase in transcriptional/translational activity and the accumulation of stored proteins and mRNA is followed by a quiescent state at maturation. A switch to a new developmental programme occurs after the pollen tube lands on the stigma with the formation of the pollen tube, growth through the style, and subsequent fertilization. Apart from the internal control mechanisms involved in this developmental programme, pollen grains must cope with physical changes during development within the anther (desiccation) and subsequently during germination on the stigma (rehydration). The metabolic and structural changes that occur throughout these processes should require signaling mechanisms to co-ordinate the appropriate response, and recent data demonstrate the presence in pollen of an array of molecules belonging to diverse signalling pathways, including mitogen-activated protein (MAP) kinases. The role of MAP kinases in pollen is discussed in the context of the various developmental and physical changes that occur throughout pollen maturation and germination.

# 1
# Pollen Development and Germination

A wide array of changes occur in pollen between microsporogenesis and attainment of the mature pollen grain. In species with bicellular pollen, the microspores undergo a single mitosis during pollen maturation, resulting in a vegetative cell and a generative cell. The latter cell divides again during pollen

---

[*] *Corresponding author.*
[1] Institute of Microbiology and Genetics, Vienna Biocenter, University of Vienna,
   Dr. Bohrgasse 9, 1030 Vienna, Austria

Results and Problems in Cell Differentiation, Vol. 27
Hirt (Ed.): MAP Kinases in Plant Signal Transduction
© Springer-Verlag Berlin Heidelberg 2000

tube growth to produce the two sperm cells. There is an increase in transcriptional/translational activity following pollen mitosis and the lumen of the vegetative cell is filled with cytoplasm containing storage materials for pollen tube formation, such as proteins, starch, lipids, sugars, as well as a large pool of inactive mRNA, and there are metabolic/structural changes to cope with desiccation of the pollen grain (Astwood and Hill 1996). Some genes are expressed throughout pollen development while others are expressed exclusively before the first pollen mitosis (early genes) or only after mitosis (late genes) (Mascarenhas 1989; Astwood and Hill 1996). The majority of the mRNAs and some or all of the proteins required for germination and early pollen tube growth are present in the pollen grain at anthesis (Mascarenhas 1989). The last phase of pollen development is initiated by dehydration. Dehydration induces a metabolically quiescent state that confers tolerance to the environmental stresses encountered by the pollen grains during dispersal by water, wind, and animals. Compounds are produced that protect the dehydrating pollen grain from desiccation damage, and water becomes increasingly bound until a stage at which the plasma membrane surrounding the pollen converts from a liquid crystalline form to a gel state. Other compounds are transported to the surface of the pollen where they mix with substances derived from the disintegrating sporophytic tapetum to form the pollen coat or tryphin.

When a pollen grain arrives on the surface of a stigma the pollen coat establishes contact and the pollen grain immediately rehydrates and swells. Within minutes a pollen tube emerges from one of the germination pores and grows rapidly. Pollen tubes are the fastest growing cells in plants. Unsaturated lipids in the pollen coat (in species with a dry stigma) or on the stigma (in species with a wet stigma) direct further tube growth by establishing a water gradient in the water/lipid emulsion on the stigma which eventually guides the tube into the stigma tissue and style (Wolters-Arts et al. 1998). Passage through the different development phases, from meiosis to microsporogenesis, pollen maturation, activation of the pollen grain on the stigma, subsequent pollen tube growth, and the interaction of the growing pollen tube with stylar tissue, and finally fertilization, require a co-ordinated regulation of metabolic and structural changes (Taylor and Hepler 1997). As in any other cell, such changes will require signal transduction mechanisms to co-ordinate the faithful enactment of the response to a given stimulus. The ability to readily isolate pollen grains provides a single cell system for studying many aspects of plant development, growth regulation, and responses to external stimuli (Vicente et al. 1991).

# 2
# Signal Transduction in Pollen

The tapetum is required for proper development of the pollen grain. Materials originating from the tapetum, either at early stages of pollen development or at later stages, when the tapetum degenerates and deposits material onto the pollen grains, may function as signaling molecules for proper pollen development, release and germination. Jasmonic acid has multiple physiological effects in plants, and has been detected in pollen (Sembdner and Parthier 1993, and references therein), although the exact origin of jasmonates in floral tissue is unknown. Jasmonic acid is able to rescue the male sterile phenotype of a maize triple mutant defective in fatty acid formation, and is suggested to have an essential role both in maturation of viable pollen and in effective dehiscence of the anther (McConn and Browse 1996). Flavonols, small molecular weight secondary plant products, are effective at nanomolar concentrations as promoters of pollen development and pollen germination (Mo et al. 1992; Ylstra et al. 1992). It seems probable that flavonols, which are essential for pollen germination, act as signaling molecules. They originate from the sporophytic cells in anthers and are transfered or absorbed by developing pollen. No free flavonols are present in developing pollen. Instead they are present as flavonol glycosides, and the aglycone form of the glycosides is generated during imbibition and germination.

Self-incompatibility (SI) responses involve signaling between pollen and the stigma. Despite the identification of changes in SI-dependent protein kinase activities and the presence of a receptor kinase at the S locus in Brassica (McCubbin and Kao 1996), direct evidence for signaling between pollen and stigma has only recently come to light. Changes in $Ca^{2+}$ concentrations (Franklin-Tong et al. 1995) and phosphorylation of a 26 kDa protein in *Papaver rhoeas* pollen (Rudd et al. 1996) are specifically induced by the SI response. A family of small, basic, cysteine-rich pollen coat proteins, one of which is known to bind to components of the S-locus, has been identified as the male determinant of self incompatibility in Brassica (Stephenson et al. 1997).

Three pollen-specific receptor-like protein kinases have been isolated, all of which contain extracellular leucine rich repeats, a transmembrane domain, and a cytoplasmic kinase domain which probably transmits the signal to intracellular effector molecules. The expression patterns differ and presumably reflect the different functions of the kinases during pollen development and pollen germination. LePRK1 and LePRK2 from tomato are expressed in late pollen development. While protein levels of LePRK1 are the same in mature and germinating pollen, the LePRK2 protein levels increase significantly after germination. LePRK2 appears to be dephosphorylated by stigma extracts, and thus the kinase may be turned off by interaction with ligands emanating from the stigma. Both of these kinases are membrane associated (Muschietti et al. 1998). PRK1 from *Petunia* is expressed a little earlier, from the mid-bicellular stage onwards (Mu et al. 1994), and appears to function during pollen development. Antisense inhibition of PRK1 leads to pollen abortion before the first mitosis (Lee et al. 1996).

Changes in $Ca^{2+}$ concentrations affect the rate and orientation of tube growth (Trewavas and Malhó 1997). Release of $Ca^{2+}$ is regulated by a phosphoinositide signal transducing system in pollen of *Papaver rhoeas* (Franklin-Tong et al. 1996). $Ca^{2+}$-dependent protein phosphorylation is also important for pollen tube growth, and is induced by the self-incompatibility response (Rudd et al. 1996). A pollen-specific calcium-dependent calmodulin-independent protein kinase (CDPK) from maize is expressed as a late gene in pollen, and has a function in pollen germination since tube growth is inhibited in the presence of antisense oligonucleotides to the kinase gene transcript (Estruch et al. 1994). Therefore, a number of elements of different signaling pathways have been identified in pollen, and functional analysis has shown how important such signaling pathways are for proper pollen development and germination.

# 3
# Changes in Pollen Grains After Hydration

The initial events that occur during rehydration of pollen and seeds show striking similarities.

1. During dehydration, the cellular membranes change from a liquid crystalline state to a gel form, and this process is then reversed upon rehydration. Part of the response of the cells during germination might be in the repair of damage caused during these processes.
2. Both pollen and seeds enter a metabolically quiescent state after dehydration, and resumption of metabolic activity is very rapid after imbibition.
3. There is a rapid shift of ribosomes into polyribosomes, and protein synthesis in the initial stages of germination uses mRNAs stored in the dry pollen/seed. This takes place even when pollen are placed in water, and the metabolic events in nondormant and dormant seeds are essentially the same after imbibition.

Much of this activity is required for normal cellular metabolism i.e. not for germination-specific events. In fact, few if any germination specific proteins have been isolated either in pollen or in seeds (Bewley 1997; Taylor and Hepler 1997). The CDPK kinase mentioned above may be one such protein in pollen.

A prominent feature of pollen activation is the reorganization of the actin cytoskeleton towards the pore where the pollen tube eventually emerges. In the dry grain of bicellular pollen, the actin network is disorganised, but in as little as 5 min after hydration a distinct reorganisation of the actin occurs, which is later directed towards the pore where the tube will eventually emerge (Tiwari and Polito 1988). In wheat pollen, the pollen grain is much less dehydrated and metabolic activity is maintained at a much higher level than in bicellular pollen (Heslop-Harrison and Heslop-Harrison 1992). Elements of the actin cytoskeleton are already focused at the germination pore at the time of pollen dispersal, so the dramatic changes in actin reorganization seen in bicellular pollen do not occur.

Another important feature of pollen activation is translation initiation. Very little transcription takes place after pollen activation, and the vast majority of proteins required for pollen germination, tube growth and fertilization have either been synthesized before rehydration or are translated from mRNAs stored in the mature pollen. In plants, two isoforms of the translation initiation factor eIF-4A exist, one of which is the pollen specific eIF-4A8. Phosphorylation of the eIF-4A has only been reported for plants, and eIF-4Ab is phosphorylated after pollen germination (op den Camp and Kuhlemeier 1998), while little phosphorylation is observed in mature pollen. Protein kinases may therefore be important for the increase in translation which occurs after pollen hydration.

# 4
# MAP Kinases in Pollen

The amplification of PCR products from pollen cDNA libraries using redundant oligonucleotides designed from conserved regions of MAP kinases suggested that MAP kinases are expressed in pollen (Wilson et al. 1995). The tobacco ntf3 MAP kinase gene appears to be expressed in pollen, as determined by Northern analysis and the ability to amplify PCR products from pollen cDNA libraries (Wilson et al. 1993, 1995), and by RT-PCR from pollen-derived mRNA (V. Voronin, C. Wilson and E. Heberle-Bors, unpubl.). However, a close homologue of ntf3 in *Petunia*, PMEK, could not be detected in the various tissues of developing anthers by in situ hybridization (Decroocq-Ferrant et al. 1995). Whether ntf3 and/or PMEK are relevant to pollen development must await further analysis.

Another tobacco MAP kinase, p45[Ntf4], is expressed in pollen (Wilson et al. 1997). Plant MAP kinase genes differ in their mode of expression. A number of plant MAP kinase genes show transcriptional induction after stress treatments such as wounding (Seo et al. 1995; Bögre et al. 1997; Zhang and Klessig 1998) or cold, drought, or salt stress (Jonak et al. 1996; Mizoguchi et al. 1996). However, the functional significance of this induction is not always clear, as it is preceded by the activation of the kinase. Ntf4 expression in pollen, by contrast, is developmentally regulated. Transcription of the ntf4 gene takes place only after the first pollen mitosis and the protein accumulates in an inactive form in the mature pollen grain. This MAP kinase is therefore expressed as a late gene in pollen. Many late expressed proteins are considered to be necessary for the activation and germination of the pollen grain. Hydration of the pollen grain, either in germination medium or in water, results in a rapid and transient activation of the MAP kinase. The kinase is inactivated before pollen tube emergence, suggesting a role in the activation of the pollen grain i.e. a return to a metabolically active state and preparation of the necessary structures and developmental programme required for tube growth. Like seed germination, some of the events beginning after rehydration of the pollen may be the same as those occurring before dehydration, while others may be specific for the re-hydration phase, eventually leading to pollen tube formation and growth. Therefore, the finding that a signal transduction component is active only transiently immediately after pollen rehydration and well before pollen tube emergence is the first indication that specific events operate in the rehydrating pollen which require activation by phosphorylation.

The actin monomer-binding protein profilin accumulates to high levels in mature pollen (Mittermann et al. 1995). It presumably has a role in the reor-ganisation of the actin network after pollen hydration (see above). However, profilin and actin do not colocalize in pollen tubes (Grote et al. 1995). Possibly, actin is released from profilactin complexes that are present in the dry pollen grain upon hydration, thus initiating actin filament assembly. Profilin itself is a potential signal transducer in this process since it can affect the pattern of phosphorylated proteins in pollen extracts (Clarke et al. 1998). Further, pollen profilin may be phosphorylated (Clarke et al. 1998). A consensus MAP kinase phosphorylation site, P-X-S/T-P (Gonzalez et al. 1991) is present at the same position in the majority of plant profilins. However, birch profilin was not phosphorylated in vitro by any of three recombinant tobacco MAP kinase pro-teins, nor by tobacco pollen extracts (C. Wilson and E. Heberle-Bors, unpubl.). Profilin does appear to be phosphorylated in mammals, but this phosphoryla-tion is by protein kinase C and $pp60^{c-src}$, a membrane-associated, non-receptor tyrosine kinase (De Corte et al. 1997; Singh et al. 1996). Phosphorylation is largely dependent on phosphoinositide binding, and in the case of protein ki-nase C, occurs on the last serine of bovine profilin (Singh et al. 1996), which is not conserved in other profilins. Phosphorylation of profilin in pollen, either by MAP kinases or any other kinase, could depend similarly on an association with other molecules, such as poly-L-proline or polyphosphoinositides (Gib-bon et al. 1998). The MAP kinase phosphorylation site is not conserved in pro-filins other than those from plants, so if this profilin site is phosphorylated by MAP kinases it would be a plant-specific control mechanism. Given the differ-ences in actin reorganisation in mature pollen and during pollen germination in dicots and monocots such as wheat (Heslop-Harrison and Heslop-Harrison 1992), it might be informative to compare the activation of the MAP kinase in wheat and in tobacco.

# 5
# The Response of MAP Kinase Signaling Pathways to Osmotic Stress

The activation of the tobacco $p45^{Ntf4}$ kinase by water alone in pollen suggested that hydration is the signal for turning on the MAP kinase cascade. The pollen grains swell rapidly upon hydration, and might activate response mechanisms similar to those identified in hypo-osmotically shocked cells from other organisms. Hypotonic cell swelling activates ion channels and ion transporters, and second messengers such as $Ca^{2+}$ and $IP_3$ (Strange 1994). Increases in $Ca^{2+}$ concentrations in plant cells after hypo-osmotic shock of tobacco cell suspension culture cells may precede the activation of protein kinases (Takahashi et al. 1997). Some of these responses may be involved in the restoration of the original cell volume, the regulatory volume decrease (RVD), through the activation of $K^+$- and $Cl^-$-selective ion channels, resulting in a net loss of salt and the efflux of water (Grinstein and Foskett 1990). Others might have functions more directly in response to increases in turgor and changes in cell membrane tension. Cardiac myocytes, which do not undergo RVD, show the activation of the MAP kinases ERK1, ERK2 and JNK1 upon hypo-osmotic-induced swelling, but not p38 (Sadoshima et al. 1996). By contrast, p38 was activated in human intestine 407 cells after hypo-osmotic cell swelling (Tilly et al. 1996). These cells do undergo RVD, but whether the differences in MAP kinase responses are due to the RVD response or whether they are cell type specific is not clear.

Cells use different MAP kinase signaling pathways in responding to hyper- or hypo-osmotic stress. In the yeasts *Saccharomyces cerevisiae* and *Schizosaccharomyces pombe*, hyperosmolarity leads to the activation of MAP kinase pathways (HOG1 and Spc1, respectively) which are involved in the accumulation of compatible solutes such as glycerol (Brewster et al. 1993; Shiozaki and Russell 1995). In *S. cerevisiae*, a different MAP kinase signaling pathway is responsive to hypo-osmotic shock (Davenport et al. 1995). This pathway comprises PKC upstream of a signaling cascade that contains the MAP kinase MPK1. The PKC-MPK1 pathway appears to have a role in restructuring of the cell wall, which might be a response to changes in membrane/cell wall tension caused by the hypo-osmolarity (Davenport et al. 1995; Kamada et al. 1995).

In plants, the *Arabidopsis* MAP kinase gene ATMPK3 was transcriptionally induced by high salinity, but also by cold and touch (Mizoguchi et al. 1996). No transcriptional induction was found after the addition of water. Within the same group of plant MAP kinases, the alfalfa MAP kinase MMK4 was activated by cold and drought, but not by high salt levels (Jonak et al. 1996). Abscisic acid, which is involved in many processes in plant physiology, including adaptation to salt and drought stress, induced the activation of a MAP kinase, of unknown identity, in barley aleurone protoplasts (Knetsch et al. 1996). The pea MAP kinase PsMAPK, which falls into the same group as p45[Ntf4], SIPK and MMK1, was able to partially complement a *hog1* mutant in yeast which might indicate a role for this kinase in a response to hyper-osmotic stress (Pöpping et al. 1996). Therefore, although there is circumstantial evidence for the involvement of MAP kinases in the response to salt stress, there are no firm data. Similarly, no data exist for hypo-osmotically-induced MAP kinase activation. Whether the activation of p45[Ntf4] after pollen hydration is a general response to hypo-osmotic stress remains to be determined.

How might the hydration of pollen grains lead to the activation of a MAP kinase pathway? Hypo-osmotically induced swelling leads to an increase in turgor and membrane tension in the cell. In cardiac myocytes (Sadoshima et al. 1996) and yeast (Kamada et al. 1995), membrane deformation caused by chlorpromazine mimics the effects of hypotonic stress, which include the activation of MAP kinases. The existence of mechanoreceptors in the membrane could be the initial response of a signaling pathway to changes in membrane tension caused by cell swelling or shrinking. One such mechano(osmo)sensor mignt be Sln1, a transmembrane dual component histidine kinase receptor upstream of the HOG1 MAP kinase in yeast in the hyperosmotic stress response (Maeda et al. 1994). Activation of such receptors might occur through a conformational change brought about by membrane stretching/deformation, leading to autophosphorylation, or association with substrates or other activating molecules (see also below).

# 6
# Specificity of the MAP Kinase Response

A single MAP kinase signaling cascade responds to a wide array of stimuli, and results in a variety of reponses. Conversely, a particular stimulus leads to the activation of different MAP kinases. How might the appropriate response to a particular stimulus be achieved? The stimulation of animal cells with growth hormones or differentiation factors have different consequences, which may be due to differences in the duration of the MAP kinase response (Marshall 1995). Different stimuli might lead to similar responses by feeding into a pathway at different levels through different upstream signaling molecules. Activation of the Spc1 MAP kinase in S. pombe by high osmolarity proceeds via the Win1 MAPKKK, but heat shock- and oxidative stress-mediated Spc1 activation are Win1 independent (Samejima et al. 1998). In yeast, most of the same elements in the signal transduction pathways involved in mating and filamentous growth are shared. Specificity may be achieved through particular elements of each pathway interacting with specific partners and by incorporation of the elements of a pathway into a multicomponent complex (Hill and Treisman 1995; Madhani and Fink 1998). Recent work in mammalian cells further demonstrates how adaptor proteins are used for specific signaling by particular MAP kinases (Elion 1998; Schaeffer et al. 1998; Whitmarsh et al. 1998).

In plants, different stimuli result in the activation of a particular MAP kinase. The tobacco MAP kinase SIPK responds to salicylic acid, elicitors, wounding, and TMV infection (Zhang and Klessig 1997, 1998; Zhang et al. 1998). MMK4 from alfalfa is responsive to wounding, touching, cold, drought, and elicitors (Bögre et al. 1996, 1997; Jonak et al. 1996; Ligterink et al. 1997). p45[Ntf4] which is most similar to SIPK from tobacco and MMK1 from alfalfa is activated after hydration of pollen. Different stresses might result in similar initial events at the cell surface or cell membrane which could activate a signaling pathway. Changes in the microtubular cytoskeleton occur through many different forms of stimulation (Nick 1998). Thus, wounding and pathogen attack lead to changes in cell wall and membrane structure. Elicitors may stimulate a response through non-specific interactions with the extracellular matrix or the binding of specific receptors. Proteins of the extracellular matrix transmit signals involved in growth, differentiation, and tissue repair (End and Engel 1991). Heterodimeric transmembrane receptors called integrins bind many components of the extracellular matrix and are important in the transmission of signals to different signaling pathways. Calreticulin binds to membrane-spanning integrin-like proteins (Rojiani et al. 1991), which have been proposed to act as mechanoreceptors (Wang et al. 1993), linking the extracellular matrix to the cytoskeleton through the binding of actin-associated proteins within focal adhesions. In tobacco cells, the phosphorylation state of a calreticulin-like protein changes in response to elicitor treatment (Droillard et al. 1997). Integrin-generated signals activate signal transduction pathways, including MAP kinase cascades (Short et al. 1998), and integrin signaling via the GTPase RhoA controls the Na-H exchanger NHE1 which is important for pH homeostasis and focal adhesion assembly (Tominaga and Barber 1998). The generation of common second messengers might be an alternative way in which different stimuli lead to the activation of the same MAP kinase. Elicitors lead to responses such as membrane depolarization, calcium influx, the release of active oxygen species, and changes in extra- and intracellular pH (Ebel and Mithöfer 1998), some of which are similar to the responses to osmotic stimulation. A common subset of responses to different stimuli could activate a single pathway, while stimulus-specific responses are elicited by other second messengers.

A recent report on the activation of MAP kinases after infection of tobacco leaves with TMV indicates that, at least in some cases, a more specific control mechanism operates. TMV infection leads to the activation of two tobacco MAP kinases, SIPK and WIPK (Zhang and Klessig 1998). The WIPK activation requires de novo synthesis of mRNA and protein while SIPK is constitutively expressed. Further, WIPK but not SIPK expression and activation is dependent on the disease resistance gene N. Therefore, while the activation profiles of the two kinases are superficially similar, the response mechanisms are clearly different with respect to TMV infection.

*Acknowledgements.* Work in the authors' laboratory was supported by the Austrian *Fonds zur Förderung der wissenschaftlichen Forschung* project P11481-GEN, the European Union BIOTECH project SIME, project number BIO4 CT96 0275, and the European Union Training and Mobility of Researchers (TMR) project RYPLOS, project number FMRX-CT96-0007.

# References

Astwood JD, Hill RD (1996) Molecular biology of male gamete development in plants – an over-
    view. In: Mohapatra SS, Knox RB (eds) Pollen Biotechnology. Chapman and Hall, New York,
    pp 3–37
Bewley JD (1997) Seed germination and dormancy. Plant Cell 9:1055–1066
Bögre L, Ligterink W, Heberle-Bors E, Hirt H (1996) Mechanosensors in plants. Nature 383:
    489–490
Bögre L, Ligterink W, Meskiene I, Barker PJ, Heberle-Bors E, Huskisson NS, Hirt H (1997)
    Wounding induces the rapid and transient activation of a specific MAP kinase pathway. Plant
    Cell 9:75–83
Brewster JL, deValoir T, Dwyer ND, Winter E, Gustin MC (1993) An osmosensing signal trans-
    duction pathway in yeast. Science 259:1760–1763
Clarke SR, Staiger CJ, Gibbon BC, Franklin-Tong VE (1998) A potential signaling role for profi-
    lin in pollen of *Papaver rhoeas*. Plant Cell 10:967–979
Davenport KR, Sohaskey M, Kamada Y, Levin DE, Gustin MC (1995) A second osmosensing sig-
    nal transduction pathway in yeast. J Biol Chem 270:30157–30161
De Corte V, Gettemans J, Vandekerckhove J (1997) Phosphatidylinositol 4,5-bisphosphate spe-
    cifically stimulates pp60c-src catalyzed phosphorylation of gelsolin and related actin-binding
    proteins. FEBS Lett 401:191–196
Decroocq-Ferrant V, Decroocq S, VanWent J, Schmidt E, Kreis M (1995) A homologue of the
    MAP/ERK family of protein kinases is expressed in vegetative and in female reproductive or-
    gans of *Petunia hybrida*. Plant Mol Biol 27:339–350
Droillard MJ, Guclu J, LeCaer JP, Mathieu Y, Guern J, Lauriere C (1997) Identification of calretic-
    ulin-like protein as one of the phosphoproteins modulated in response to oligogalacturonides
    in tobacco cells. Planta 202:341–348
Ebel J, Mithöfer A (1998) Early events in the elicitation of plant defense. Planta 206:335–348
Elion E (1998) Routing MAP kinase cascades. Science 281:1625–1626
End P, Engel J (1991) Multidomain proteins of the extracellular matrix and cellular growth. In:
    McDonald JA, Mecham RP (eds) Receptors for extracellular matrix. Academic Press, New
    York, pp 79–129
Estruch J, Kadwell S, Merlin E, Crossland L (1994) Cloning and characterization of a maize pol-
    len-specific calcium-dependent calmodulin-independent protein kinase. Proc Natl Acad Sci
    USA 91:8837–8841
Franklin-Tong VE, Ride J, Franklin FCH (1995) Recombinant stigmatic self-incompatibility (S-)
    protein elicits a $Ca^{2+}$ transient in pollen of *Papaver rhoeas*. Plant J 8:299–307
Franklin-Tong VE, Drøbak BK, Allan AC, Watkins PAC, Trewavas AJ (1996) Growth of pollen
    tubes of *Papaver rhoeas* is regulated by a slow-moving calcium wave propagated by inositol
    1,4,5-trisphosphate. Plant Cell 8:1305–1321
Gibbon BC, Zonia LE, Kovar DR, Hussey PJ, Staiger CJ (1998) Pollen profilin function depends
    on interaction with proline-rich motifs. Plant Cell 10:981–993
Gonzalez FA, Raden DL, Davis RJ (1991) Identification of substrate recognition determinants for
    human ERK1 and ERK2 protein kinases. J Biol Chem 266:22159–22163
Grinstein S, Foskett JK (1990) Ionic mechanisms of cell volume regulation in leukocytes. Annu
    Rev Physiol 52:399–414
Grote M, Swoboda I, Meagher RB, Valenta R (1995) Localization of profilin- and actin-like im-
    munoreactivity in in vitro-germinated tobacco pollen tubes by electron microscopy after spe-
    cial water-free fixation techniques. Sex Plant Reprod 8:180–186
Heslop-Harrison J, Heslop-Harrison Y (1992) Intracellular motility, the actin cytoskeleton and
    germinability in the pollen of wheat (*Triticum aestivum* L.). Sex Plant Reprod 5:247–255
Hill CS, Treisman R (1995) Transcriptional regulation by extracellular signals: mechanisms and
    specificity. Cell 80:199–211

Jonak C, Kiegerl S, Ligterink W, Barker PJ, Huskisson NS, Hirt H (1996) Stress signalling in plants: a mitogen-activated protein kinase pathway is activated by cold and drought. Proc Natl Acad Sci USA 93:11 274–11 279

Kamada Y, Jung US, Piotrowski J, Levin DE (1995) The protein kinase C-activated MAP kinase pathway of *Saccharomyces cerevisiae* mediates a novel aspect of the heat shock response. Gene Dev 9:1559–1571

Lee HS, Karunanandaa B, McCubbin A, Gilroy S, Kao TH (1996) PRK1, a receptor-like kinase of *Petunia inflata*, is essential for postmeiotic development of pollen. Plant J 9:613–624

Ligterink W, Kroj T, Zur Nieden U, Hirt H, Scheel D (1997) Receptor-mediated activation of a MAP kinase in pathogen defense of plants. Science 276:2054–2057

Madhani HD, Fink GR (1998) The riddle of MAP kinase signaling specificity. Trends Genet 14:151–155

Maeda T,Wurgler-Murphy SM, Saito H (1994) A two-component system that regulates an osmo-sensing MAP kinase cascade in yeast. Nature 369:242–245

Marshall CJ (1995) Specificity of receptor tyrosine kinase signaling: transient versus sustained extracellular signal-regulated kinase activation. Cell 80:179–185

Mascarenhas JP (1989) The male gametophyte of flowering plants. Plant Cell 1:657–664

McConn M, Browse J (1996) The critical requirement for linolenic acid is pollen development, not photosynthesis, in an Arabidopsis mutant. Plant Cell 8:403–416

McCubbin AG, Kao T (1996) Molecular mechanisms of self-incompatibility. Curr Opin Biotechnol 7:150–154

Mittermann I, Swoboda I, Pierson E, Eller N, Kraft D, Valenta R, Heberle-Bors E (1995) Molecular cloning and characterization of profilin from tobacco *(Nicotiana tabacum)*: increased profilin expression during pollen maturation. Plant Mol Biol 27:137–146

Mizoguchi T, Irie K, Hirayama T, Hayashida N, Yamaguchi Shinozaki K, Matsumoto K, Shinozaki K (1996) A gene encoding a mitogen-activated protein kinase kinase kinase is induced simultaneously with genes for a mitogen-activated protein kinase and an S6 ribosomal protein kinase by touch, cold and water stress in *Arabidopsis thaliana*. Proc Natl Acad Sci USA 93:765–769

Mo Y, Nagel C, Taylor LP (1992) Biochemical complementation of chalcone synthase mutants defines a role for flavanols in functional pollen. Proc Natl Acad Sci USA 89:7213–7217

Mu JH, Lee HS, Koa TH (1994) Characterization of a pollen-expressed receptor-like kinase gene of *Petunia inflata* and the activity of its encoded kinase. Plant Cell 6:709–721

Muschietti J, Eyal Y, McCormick S (1998) Pollen tube localization implies a role in pollen-pistil interactions for the tomato receptor-like protein kinases LePRK1 and LePRK2. Plant Cell 10:319–330

Nick P (1998) Signaling to the microtubular cytoskeleton in plants. Int Rev Cytol 184:33–80

op den Camp RGL, Kuhlemeier C (1998) Phosphorylation of tobacco eukaryotic translation initiation factor 4A upon pollen tube germination. Nucleic Acids Res 26:2058–2062

Pöpping B, Gibbons T, Watson MD (1996) The *Pisum sativum* MAP kinase homologue (Ps-MAPK) rescues the *Saccharomyces cerevisiae* hog1 mutant under conditions of high osmotic stress. Plant Mol Biol 31:355–363

Rojiani MV, Finlay BB, Gray V, Dedhar S (1991) In vitro interaction of a polypeptide homologous to human Ro/SS-A antigen (calreticulin) with the highly conserved amino acid sequence in the cytoplasmic domain of integrin a subunits. Biochemistry 30:9859–9866

Rudd JJ, Franklin CH, Lord JM, Franklin-Tong VE (1996) Increased phosphorylation of a 26-kD pollen protein is induced by the self-incompatibility response in *Papaver rhoeas*. Plant Cell 8:713–724

Sadoshima J, Qiu Z, Morgen JP, Izumo S (1996) Tyrosine kinase activation is an immediate and essential step in hypotonic cell swelling-induced ERK activation and c-*fos* gene expression in cardiac myocytes. EMBO J 15:5535–5546

Samejima I, Mackie S, Warbrick E, Weisman R, Fantes PA (1998) The fission yeast mitotic regulator *win1*⁺ encodes an MAP kinase kinase kinase that phosphorylates and activates Wis1 MAP kinase kinase in response to high osmolarity. Mol Biol Cell 9:2325–2335

Schaeffer HJ, Catling AD, Eblen ST, Collier LS, Krauss A, Weber MJ (1998) MP1: a MEK binding partner that enhances enzymatic activation of the MAP kinase cascade. Science 281: 1668–1671

Sembdner G, Parthier B (1993) The biochemistry and the physiological and molecular actions of jasmonates. Annu Rev Plant Physiol Plant Mol Biol 44:569–589

Seo S, Okamoto M, Seto H, Ishizuka K, Sano H, Ohashi Y (1995) Tobacco MAP kinase: a possible mediator in wound signal transduction pathways. Science 270:1988–1992

Shiozaki K, Russell P (1995) Cell-cycle control linked to extracellular environment by MAP kinase pathway in fission yeast. Nature 378:739–743

Short SM, Talbott GA, Juliano RL (1998) Integrin-mediated signaling events in human endothelial cells. Mol Biol Cell 9:1969–1980

Singh SS, Chauhan A, Murakami N, Styles J, Elzinga M, Chauhan VP (1996) Phosphoinositide-dependent in vitro phosphorylation of profilin by protein kinase C. Phospholipid specificity and localization of the phosphorylation site. Receptors Signal Transduct 6:77–86

Stephenson AG, Doughty J, Dixon S, Elleman C, Hiscock S, Dickinson HG (1997) The male determinant of self-incompatibility in *Brassica oleracea* is located in the pollen coating. Plant J 12:1351–1359

Strange K (1994) Cellular and molecular physiology of cell volume regulation. CRC Press, Boca Raton, Florida

Takahashi K, Isobe M, Muto S (1997) An increase in cytosolic calcium ion concentration precedes hypoosmotic shock-induced activation of protein kinases in tobacco cell suspension culture cells. FEBS Lett 401:202–206

Taylor LP, Hepler PK (1997) Pollen germination and tube growth. Annu Rev Plant Physiol Plant Mol Biol 48:461–491

Tilly BC, Gaestel M, Engel K, Edixhoven MJ, de Jonge HR (1996) Hypo-osmotic cell swelling activates the p38 MAP kinase signalling cascade. FEBS Lett 395:133–136

Tiwari SC, Polito VS (1988) Spatial and temporal organization of actin during hydration, activation, and germination of pollen in *Pyrus communis* L.: a population study. Protoplasma 147: 5–15

Tominaga T, Barber DL (1998) Na-H exchange acts downstream of RhoA to regulate integrin-induced cell adhesion and spreading. Mol Biol Cell 9:2287–2303

Traverse S, Gomez N, Paterson H, Marshall C, Cohen P (1992) Sustained activation of the mitogen-activated protein (MAP) kinase cascade may be required for differentiation of PC12 cells. Comparison of the effects of nerve growth factor and epidermal growth factor. Biochem J 288:351–355

Trewavas AJ, Malhó R (1997) Signal perception and transduction: the origin of the phenotype. Plant Cell 9:1181–1195

Vicente O, Benito Moreno RM, Heberle-Bors E (1991) Pollen cultures as a tool to study plant development. Cell Biol Rev 25:295–305

Wang N, Butler JP, Ingber DE (1993) Mechanotransduction across the cell surface and through the cytoskeleton. Science 260:1124–1127

Whitmarsh AJ, Cavanagh J, Tournier C, Yasuda J, Davis RJ (1998) A mammalian scaffold complex that selectively mediates MAP kinase activation. Science 281:1671–1674

Wilson C, Eller N, Gartner A, Vicente O, Heberle-Bors E (1993) Isolation and characterization of a tobacco cDNA clone encoding a putative MAP kinase. Plant Mol Biol 23:543–551

Wilson C, Anglmayer R, Vicente O, Heberle-Bors E (1995) Molecular cloning, functional expression in *Escherichia coli*, and characterization of multiple mitogen- activated-protein kinases from tobacco. Eur J Biochem 233:249–257

Wilson C, Voronin V, Touraev A, Vicente O, Heberle-Bors E (1997) A developmentally regulated MAP kinase activated by hydration in tobacco pollen. Plant Cell 9:2093–2100

Wolters-Arts M, Lush WM, Mariani C (1998) Lipids are required for directional pollen-tube growth. Nature 392:818–821

Ylstra B, Touraev A, Benito Moreno RM, Stöger E, van Tunen AJ, Vicente O, Mol JNM, Heberle-Bors E (1992) Flavonols stimulate development, germination, and tube growth of tobacco pollen. Plant Physiol 100:902–907

Zhang S, Klessig DF (1997) Salicylic acid activates a 48-kDa MAP kinase in tobacco. Plant Cell 9:809–824

Zhang S, Klessig DF (1998) Resistance gene N-mediated de novo synthesis and activation of a tobacco mitogen-activated protein kinase by tobacco mosaic virus infection. Proc Natl Acad Sci USA 95:7433–7438

Zhang S, Du H, Klessig DF (1998) Activation of the tobacco SIP kinase by both a cell wall-derived carbohydrate elicitor and purified proteinaceous elicitins from *Phytophthora* spp. Plant Cell 10:435–449

# Mitogen-Activated Protein Kinases and Wound Stress

Shigemi Seo and Yuko Ohashi[1]

**Summary:** One of the most severe environmental stresses that plants encounter during their life cycle is wounding. Plants respond to wound stress by activating a set of genes that encode proteins involved in healing injured tissues. In recent years, mitogen-activated protein kinases have been implicated to be key signal molecules in the initial signal transduction pathways that mediate this stress to expression of genes.

## 1
## Introduction

Plants cope with wound stress such as physical injury and herbivore or insect attack by activating a set of cell cycle- and defense-related genes. These gene products such as cdc2 (Hemerly et al. 1993), phenylalanine ammonia lyase (Lawton and Lamb 1987), proteinase inhibitors (PIs; Ryan 1990), and basic pathogenesis-related (PR) proteins (Memelink et al. 1990; Brederode et al. 1991) are mainly involved in wound healing of damaged tissues and in protection against pathogen infection or insect proteases. In many cases, activation of these genes occurs not only in wounded tissues but also at unwounded sites. Hence, wound response is a good model system for studies of systemic responses of plants to external stimuli.

In recent years, much effort has been directed to the elucidation of the initial signal transduction pathways that mediate wound stress on the expression of defense-related genes. Jasmonic acid and its methyl ester accumulate in response to wound stress (Creelman et al. 1992; Albrecht et al. 1993) and induce expression of many wound-inducible genes (Staswick 1992; Reinbothe et al. 1994). Thus, jasmonates have been proposed as key signal molecules in wound signal transduction pathways (Koiwa et al. 1997; Seo et al. 1997; Wasternack and Parthier 1997). The phytohormones ethylene (O'Donnell et al. 1996) and

---

[1] Department of Molecular Genetics, National Institute of Agrobiological Resources, Tsukuba, Ibaraki 305-8602, Japan

Results and Problems in Cell Differentiation, Vol. 27
Hirt (Ed.): MAP Kinases in Plant Signal Transduction
© Springer-Verlag Berlin Heidelberg 2000

abscisic acid (ABA; Peña-Cortés et al. 1995) have also been considered to facilitate signal transduction as signal molecules. As factors that transmit wound signal to jasmonates, systemin (Bergey et al. 1996) and electrical currents (Wildon et al. 1992) have been proposed. However, very little is known about the molecular mechanisms of the initial response immediately after perception of wound stress of plants. Recently, activation of mitogen-activated protein kinases (MAPKs) in response to wound stress has been found in various plant species such as tobacco, alfalfa, tomato, and *Arabidopsis thaliana*, suggesting involvement of MAPKs in wound signal transduction.

MAPKs are a family of serine/threonine protein kinases that have been well-conserved in all eukaryotes through evolution and are activated upon phosphorylation of threonine and tyrosine residues within the TXY motif in kinase subdomain VIII (Pelech and Sanghera 1992; Nishida and Gotoh 1993; Jonak et al. 1994; Nishihama et al. 1995). In animals and yeasts, MAPKs function in transduction of extracellular stimuli into intracellular signals to regulate expression of genes that are essential for a wide variety of cellular processes, including cell proliferation, differentiation, and stress responses (Blumer and Johnson 1994; Davis 1994; Cano and Mahadevan 1995; Herskowitz 1995). Plant MAPKs have also been implicated in defense and phytohormone responses (Suzuki and Shinshi 1996; Hirt 1997; Machida et al. 1997; Mizoguchi et al. 1997). In this chapter, we review what is currently known about the roles of MAPKs in wound signal transduction.

## 2
## Evidence for Activation of MAPK-Like Protein Kinases by Wounding

Reversible phosphorylation of proteins plays an important role in the transduction of external signals. Protein kinases and protein phosphatases responsible for such phosphorylation are, therefore, cardinal factors in signal transduction pathways. Although many genes encoding protein kinases and protein phosphatases have been identified in various plant species (Stone and Walker, 1995; Redhead and Palme, 1996), little is known about their physiological functions. The involvement of protein phosphorylation/dephosphorylation in plant defense responses has been reported in several plant species such as tomato (Farmer et al. 1989,1991; Felix et al. 1991), soybean (MacKintosh et al. 1994; Chandra and Low 1995), and parsley (Dietrich et al. 1990). Direct evidence for activation of protein kinases by wound stress has come from the observation that kinase activity increased after cutting of tobacco leaves using an in-gel kinase assay with myelin basic protein (MBP) as a substrate (Usami et al. 1995). The increase in the kinase activity on cutting was rapid and transient. The following biochemical features of the tobacco wound-activated protein kinase suggested that it belongs to the MAPK family: (1) an apparent molecular mass of 46 kDa which is within the range of the family (40–50 kDa), and (2) phosphorylation of MBP which is a substrate of MAPKs. In a separate study, Seo et al. (1995) detected an increase in a 46-kDa MBP kinase activity in wounded leaves of tobacco plants. These studies suggested that MAPKs are involved in the transduction of wound signal.

# 3
# Evidence for Activation of MAPKs by Wounding

Direct evidence for activation of specific MAPKs in response to wound stress was obtained with a specific antibody for an alfalfa MAPK. In alfalfa, three MAPK homologs have been identified; Msk7 [Jonak et al. 1993; also referred to as MsERK1 (Duerr et al. 1993)], MMK2 (Jonak et al. 1995), and MMK4 (Jonak et al. 1996). Bögre et al. (1997) measured MBP kinase activity in immunoprecipitates from extracts of wounded leaves with an antibody against a synthetic peptide corresponding to the C-terminal ten amino acids of MMK4. Rapid and transient increases in MBP kinase activity were observed in the antibody-immunoprecipitate complex. JA, ABA, and an electrical current were unable to activate MMK4, suggesting that the MAPK activation pathway acts independently of these signal molecules, or that MAPK functions upstream of them.

In tobacco, five MAPK homologs have been identified; Ntf3 (Wilson et al. 1993), Ntf4, Ntf6 (Wilson et al. 1995), WIPK (Seo et al. 1995), and SIPK (Zhang and Klessig 1997). WIPK (wound-induced protein kinase) was identified as the protein encoded by a cDNA whose transcript accumulates in response to wounding. Zhang and Klessig (1998) demonstrated that SIPK, which was originally identified as salicylic acid (SA)-induced protein kinase, is activated by wounding using a specific antibody. The activation accompanied phosphorylation on the tyrosine residue. The authors have also reported that no enzymatic activation of WIPK after wounding is detected with the kinase-specific antibody. However, we recently found that WIPK is actually activated by wounding, using an antibody against a synthetic peptide identical to the antigen peptide described in the 1998 report by Zhang and Klessig (1998) (Seo et al. 1999). The reason for the different results in the two laboratories is not clear.

In animals and yeasts, the MAP kinase cascade is based on phosphorylation and dephosphorylation of the constituent kinases, and the dephosphorylation of MAP kinases is mediated by the kinase-specific serine/threonine or tyrosine phosphatases (Sun et al. 1993; Doi et al. 1994; Shiozaki and Russell 1995; Wurgler-Murphy et al. 1997). Therefore, such phosphatases function as negative regulators of the MAP kinase cascade. Involvement of phosphatases in MAPK pathways in plants was demonstrated in wound responses of alfalfa (Meskiene et al. 1998).

## 4
## Transcriptional Activation of MAPKs by Wounding

In animals and yeasts, activation by external stimuli of MAPKs occurs at the post-translational level. Also in plants, post-translational activation of MAPKs in response to external stimuli has been reported in MMK4 from alfalfa (Jonak et al. 1996; Bögre et al. 1997) and SIPK (Zhang et al. 1998; Zhang and Klessig 1998) and Ntf4 (Wilson et al. 1997) from tobacco. Interestingly, certain MAPKs in plants are activated at the transcriptional level. For example, the transcript

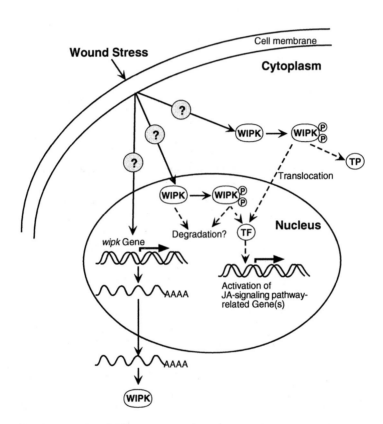

**Fig. 1.** Proposed model for activation of WIPK upon wounding. Wound signal(s) are transmitted to cytoplasmic WIPK probably via mediators. Some of the activated WIPK phosphorylate target protein(s) *TP* in the cytoplasm, and some of the activated proteins are translocated to the nucleus and activate transcription factor(s) *TF* of jasmonate-signaling pathway-related gene(s). It is also possible that nuclear WIPK is activated. After wounding, the WIPK protein level in the nucleus is decreased by an unknown mechanism. Wound signal(s) is also transmitted to the *wipk* gene, and the mRNA begins to accumulate to compensate for a change in the steady state level of WIPK protein. Information on SIPK is omitted because the relationship between it and WIPK in wound responses of tobacco plants is unclear

for WIPK accumulates after wounding of tobacco plants (Seo et al. 1995). The alfalfa MMK4, which shows high sequence homology to WIPK, is also transcriptionally activated in response to wounding (Bögre et al. 1997). In addition, transcripts for the *Arabidopsis* ATMPK3 (Mizoguchi et al. 1996) and the parsley ERMK (Ligterink et al. 1997), which are very similar to WIPK, accumulate in response to drought, cold, touch, or elicitor. In contrast to these MAPKs, activation of the tobacco SIPK by SA, elicitor, and wounding occurs at the posttranslational level but not at the transcriptional level (Zhang et al. 1998; Zhang and Klessig 1998).

What is the physiological meaning of the transcriptional activation of MAPKs in plants? Recently, we found that WIPK exists not only in the cytoplasm but also in the nucleus of tobacco plant cells (S. Seo and Y. Ohashi., unpubl. data). Post-translational activation of WIPK after wounding was detected in both cytoplasmic and nuclear fractions. These results suggest that, upon wounding, cytoplasmic WIPK is activated, and some of the activated proteins are translocated to the nucleus. Another possibility is that both the cytoplasmic and nuclear WIPK are activated. These activated WIPK may phosphorylate transcription factors of defense-related genes. To our surprise, it was found that, upon wounding, the protein level in the cytoplasmic fraction is of steady state while that in the nuclear fraction decreases transiently. Based on this observation, we speculate the following regarding transcriptional activation of WIPK. After completion of their role in the nucleus, phosphorylated WIPK proteins are degraded by unknown mechanisms. Assuming that a steady state level of WIPK proteins is necessary in order to respond to unpredictable wound stress, *wipk* transcript must accumulate to compensate for any change in the level. A possible model for activation by wounding of WIPK in tobacco cells is illustrated in Fig. 1. Obviously, further studies on the mechanisms of the protein decrease of WIPK will be required.

# 5
# Systemic Activation of MAPKs by Wounding

Activation of MAPKs by wounding occurs systemically. It has been reported that, when the stem of a tobacco plant was cut, accumulation of transcript for WIPK was observed in unwounded upper leaves of the same plant (Seo et al. 1995). A systemic response by the *wipk* gene after wounding was found to occur at the post-translational level (Seo et al. 1999). Also, in wounded tomato plants, systemic activation of a 48-kDa MBP-phosphorylating protein kinase has been found (Stratmann and Ryan 1997).

# 6
# What Is the Initial Wound Signal for Activation of MAPKs?

Activation of MAPKs after wounding occurs within a few minutes (Seo et al. 1995; Usami et al. 1995; Bögre et al. 1997; Zhang and Klessig 1998). In addition, this rapid response of MAPKs occurs systemically. What signal(s) triggers such a rapid local and systemic activation of MAPKs? In plant cells, wounding induces a rapid change in the electrical current (Wildon et al. 1992) and instantaneous change of hydraulic pressure in the whole plant body (Malone 1992). Such physical changes could locally and systemically activate MAPK pathways. However, there has been a report that electrical current did not activate MMK4 in alfalfa (Bögre et al. 1997). Identification of the initial wound signal for activation of MAPKs is expected.

# 7
# Studies of Function of MAPKs Using Transgenic Plants

Despite many reports on isolation and identification of MAPK homologs in plants, the physiological functions are not fully understood. Research using transgenic plants is one means of investigation. To study the role of WIPK in wound responses, transgenic tobacco plants for overexpression of the gene were generated. In transformants, suppression of the endogenous *wipk* gene and overproduction of transcripts for introduced *wipk* gene were observed (Seo et al. 1995). Unexpectedly, in the transformants, wound-induced transient increase in MBP kinase activity, including WIPK activity, was inhibited. Such plants did not accumulate either jasmonates or transcripts for jasmonate-inducible genes encoding PI-II and basic PR-1 protein (Fig. 2). These results suggested that WIPK functions as a positive regulator of jasmonate synthesis. The involvement of protein kinase(s) in the synthesis of jasmonates has been pointed out (Farmer and Ryan 1992). It is highly likely that WIPK is one of these protein kinases.

Interestingly, in the *wipk*-silenced transgenic plants, both SA and transcripts for SA-inducible genes encoding, for example, acidic PR proteins were found to accumulate after wounding (Seo et al. 1995). In general, wounding does not cause accumulation of SA (Malamy et al. 1990) and SA-inducible acidic PR proteins (Memelink et al. 1990; Brederode et al. 1991). Why did SA accumulate in the wounded transgenic plants? SA and its derivatives inhibit the synthesis of jasmonates (Peña-Cortés et al. 1993) and jasmonate-inducible proteins (Doares et al. 1995). On the other hand, jasmonates inhibit the accumulation of SA and SA-inducible acidic PR proteins (Sano et al. 1996; Niki et al. 1998). The antagonistic effect of jasmonates and SA could explain the abnormal SA accumulation in the *wipk*-silenced plants. Namely, the SA accumula-

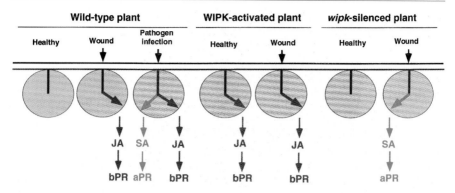

**Fig. 2.** Proposed signal flows on wounding in wild-type and WIPK-activated and *wipk*-silenced transgenic plants. Activated signal flows are indicated by *arrows*. *Circles* represent WIPK as a signal switch which directs signals to jasmonate *JA* and *SA* channels. *bPR* and *aPR* indicate jasmonate-inducible basic PR proteins and SA-inducible acidic PR proteins, respectively. In loss-of-function transgenic plants (*wipk*-silenced plant), wound signal does not flow to JA channel. In gain-of-function transgenic plants (WIPK-activated plants), the JA-signaling pathway is activated constitutively

tion was triggered by a low level of jasmonates. Thus, WIPK may also regulate SA-signaling pathways (Fig. 2).

Since the *wipk*-silenced transgenic plants are loss-of-function transgenic plants, the data obtained by analyses using such plants are only indirect evidence for the role of WIPK in wound response. Therefore, selection of gain-of-function transgenic plants was performed in our laboratory. If WIPK functions as a positive regulator for jasmonate synthesis, the gain-of-function plants would show constitutive accumulation of jasmonates and transcript for *PI-II* gene. Based on this idea, plants that showed such a phenotype were selected in about 30 lines of transformants expressing the sense *wipk* gene under the control of a constitutive promoter. In healthy leaves, three lines were found to contain about threefold higher levels of jasmonates than the wild-type plants and show constitutive accumulation of *PI-II* transcript (Seo et al. 1999). In these lines, enzymatic activation of WIPK was found constitutively. It was concluded that the three lines are gain-of-function transgenic plants. Thus, analyses using two types of transgenic plants strongly suggest that enzymatic activation of WIPK is required for synthesis of jasmonates in tobacco plants (Fig. 2).

In tobacco, at least two MAPKs, WIPK and SIPK, are activated in response to wounding. What differences are found in their functions in wound response? SIPK seems to be a multistress-responsive MAPK because it is activated in response to various stimuli such as wounding, SA and elicitor (Zhang and Klessig 1997; Zhang and Klessig 1998). Although it has been reported that about a one order of magnitude higher amount of SIPK protein than WIPK protein was found in tobacco leaves (Zhang and Klessig 1998), the physiological function of SIPK in defense responses is not yet clear. Further analyses of WIPK and SIPK will help in the understanding of wound signal transduction pathways.

# 8
# Concluding Remarks

Recent studies have demonstrated that MAPKs are involved in the transduction of wound signals in plants. However, little is known about their roles in wound response. To develop a better understanding, we must focus on: (1) identification of upstream and downstream factors of MAPKs, (2) identification of wound signal(s) for rapid activation of MAPKs, (3) elucidation of the physiological importance of transcriptional activation of MAPKs, and (4) elucidation of phosphorylation/dephosphorylation mechanisms of MAPKs. Analyses with transgenic plants, a yeast two-hybrid system and cell manipulation techniques will aid in such studies.

# References

Albrecht T, Kehlen A, Stahl K, Knöfel H-D, Sembdner G, Weiler EW (1993) Quantification of rapid, transient increases in jasmonic acid in wounded plants using a monoclonal antibody. Planta 191:86–94

Bergey DR, Howe GA, Ryan CA (1996) Polypeptide signaling for plant defensive genes exhibits analogies to defense signaling in animals. Proc Natl Acad Sci USA 93:12053–12058

Blumer KJ, Johnson GL (1994) Diversity in function and regulation of MAP kinase pathways. Trends Biochem Sci 19:236–240

Bögre L, Ligterink W, Meskiene I, Barker PJ, Heberle-Bors E, Huskisson NS, Hirt H (1997) Wounding induces the rapid and transient activation of a specific MAP kinase pathway. Plant Cell 9:75–83

Brederode FTh, Linthorst HJM, Bol JF (1991) Differential induction of acquired resistance and PR gene expression in tobacco by virus infection, ethephone treatment, UV light and wounding. Plant Mol Biol 17:1117–1125

Cano E, Mahadevan LC (1995) Parallel signal processing among mammalian MAPKs. Trends Biochem Sci 20:117–122

Chandra S, Low PS (1995) Role of phosphorylation in elicitation of the oxidative burst in cultured soybean cells. Proc Natl Acad Sci USA 92:4120–4123

Cobb MH, Goldsmith EJ (1995) How MAP kinases are regulated. J Biol Chem 270:14843–14846

Creelman RA, Tierney ML, Mullet JE (1992) Jasmonic acid/methyl jasmonate accumulate in wounded soybean hypocotyls and modulate wound gene expression. Proc Natl Acad Sci USA 89:4938–4941

Davis RJ (1994) MAPKs: new JNK expands the group. Trends Biochem Sci 19:470–473

Dietrich A, Mayer JE, Hahlbrock K (1990) Fungal elicitor triggers rapid, transient, and specific protein phosphorylation in parsley cell suspension cultures. J Biol Chem 265:6360–6368

Doares SH, Narvez-Vsquez J, Conconi A, Ryan CA (1995) Salicylic acid inhibits synthesis of proteinase inhibitors in tomato leaves induced by systemin and jasmonic acid. Plant Physiol 108:1741–1746

Doi K, Gartner A, Ammerer G, Errede B, Shinkawa H, Sugimoto K, Matsumoto K (1994) MSG5, a novel protein phosphatase promotes adaptation to pheromone response in S. cerevisiae. EMBO J 13:61–70

Duerr B, Gawienowski M, Ropp T, Jacobs T (1993) MsERK1: a mitogen-activated protein kinase from a flowering plant. Plant Cell 5:87–96

Farmer EE, Ryan CA (1992) Octadecanoid precursors of jasmonic acid activate the synthesis of wound-inducible proteinase inhibitors. Plant Cell 4:129–134

Farmer EE, Pearce G, Ryan CA (1989) In vitro phosphorylation of plant plasma membrane proteins in response to the proteinase inhibitor inducing factor. Proc Natl Acad Sci USA 86: 1539–1542

Farmer EE, Moloshok TD, Saxton MJ, Ryan CA (1991) Oligosaccharide signaling in plants. Specificity of oligouronide-enhanced plasma membrane protein phosphorylation. J Biol Chem 266:3140–3145

Felix G, Grosskopf DG, Regenass M, Boller T (1991) Rapid changes of protein phosphorylation are involved in transduction of the elicitor signal in plant cells. Proc Natl Acad Sci USA 88: 8831–8834

Hemerly AS, Ferreira P, Engler JdeA, Van Montagu M, Engler G, Inzé D (1993) cdc2a expression in Arabidopsis is linked with competence for cell division. Plant Cell 5:1711–1723

Herskowitz I (1995) MAP kinase pathways in yeast: for mating and more. Cell 80:187–197

Hirt H (1997) Multiple roles of MAP kinases in plant signal transduction. Trends Plant Sci 2: 11–15

Jonak C, Heberle-Bors E, Hirt H (1994) MAP kinases: universal multi-purpose signaling tools. Plant Mol Biol 24:407–416

Jonak C, Kiegerl S, Lloyd C, Chan J, Hirt H (1995) MMK2, a novel alfalfa MAP kinase, specifically complements the yeast MPK1 function. Mol Gen Genet 248:686–694

Jonak C, Kiegerl S, Ligterink W, Barker PJ, Huskisson NS, Hirt H (1996) Stress signaling in plants: a mitogen-activated protein kinase pathway is activated by cold and drought. Proc Natl Acad Sci USA 93:11274–11279

Koiwa H, Bressan RA, Hasegawa PM (1997) Regulation of protease inhibitors and plant defense. Trends Plant Sci 2:379–384

Lawton MA, Lamb CJ (1987) Transcriptional activation of plant defense genes by fungal elicitor, wounding, and infection. Mol Cell Biol 7:335–341

Ligterink W, Kroj T, Zur Nieden U, Hirt H, Scheel D (1997) Receptor-mediated activation of a MAP kinase in pathogen defense of plants. Scince 276:2054–2057

Machida Y, Nishihama R, Kitakura S (1997) Progress in studies of plant homologs of mitogen-activated protein (MAP) kinase and potential upstream components in kinase cascades. Crit Rev Plant Sci 16:481–496

MacKintosh C, Lyon GD, MacKintosh RW (1994) Protein phosphatase inhibitors activate antifungal defence responses of soybean cotyledons and cell cultures. Plant J 5:137–147

Malamy J, Carr JP, Klessig DF, Raskin I (1990) Salicylic acid: a likely endogenous signal in the resistance response of tobacco to viral infection. Science 250:1002–1004

Malone M (1992) Kinetics of wound-induced hydraulic signals and variation potentials in wheat seedlings. Planta 187:505–510

Memelink J, Linthorst HJM, Schilperoort RA, Hoge JHC (1990) Tobacco genes encoding acidic and basic isoforms of pathogenesis-related proteins display different expression patterns. Plant Mol Biol 14:119–126

Meskiene I, Bögre L, Glaser W, Balog J, Brandstötter M, Zwerger K, Ammerer G, Hirt H (1998) MP2C, a plant protein phosphatase 2C, functions as a negative regulator of mitogen-activated protein kinase pathways in yeast and plants. Proc Natl Acad Sci USA 95:1938–1943

Mizoguchi T, Irie K, Hirayama T, Hayashida N, Yamaguchi-Shinozaki K, Matsumoto K, Shinozaki K (1996) A gene encoding a mitogen-activated protein kinase kinase kinase is induced simultaneously with genes for a mitogen-activated protein kinase and an S6 ribosomal protein kinase by touch, cold, and water stress in Arabidopsis thaliana. Proc Natl Acad Sci USA 93: 765–769

Mizoguchi T, Ichimura K, Shinozaki K (1997) Environmental stress response in plants: the role of mitogen-activated protein kinases. Trends Biotechnol 15:15–19

Niki T, Mitsuhara I, Seo S, Ohtsubo N, Ohashi Y (1998) Antagonistic effect of salicylic acid and jasmonic acid on the expression of pathogenesis-related (PR) protein genes in wounded mature tobacco leaves. Plant Cell Physiol 39:500–507

Nishida E, Gotoh Y (1993) The MAP kinase cascade is essential for diverse signal transduction pathways. Trends Biochem Sci 18:128–131

Nishihama R, Banno H, Shibata W, Hirano K, Nakashima M, Usami S, Machida Y (1995) Plant homologues of components of MAPK (mitogen-activated protein kinase) signal pathways in yeast and animal cells. Plant Cell Physiol 36:749–757

O'Donnell PJ, Calvert C, Atzorn Z, Wasternack C, Leyser HMO, Bowles DJ (1996) Ethylene as a signal mediating the wound response of tomato plants. Science 274:1914–1917

Pelech SL, Sanghera JS (1992) Mitogen-activated protein kinases: versatile transducers for cell signaling. Trends Biochem Sci 17:233–238

Peña-Cortés H, Albrecht T, Prat S, Weiler EW, Willmitzer L (1993) Aspirin prevents wound-induced gene expression in tomato leaves by blocking jasmonic acid biosynthesis. Planta 191:123–128

Peña-Cortés H, Fisahn J, Willmitzer L (1995) Signals involved in wound-induced proteinase inhibitor II gene expression in tomato and potato plants. Proc Natl Acad Sci USA 92:4106–4113

Redhead CR, Palme K (1996) The genes of plant signal transduction. Crit Rev Plant Sci 15:425–454

Reinbothe S, Mollenhauer B, Reinbothe C (1994) JIPs and RIPs: the regulation of plant gene expression by jasmonates in response to environmental cues and pathogens. Plant Cell 6:1197–1209

Ryan CA (1990) Proteinase inhibitors in plants: genes for improving defenses against insects and pathogens. Annu Rev Phytopathol 28:425–449

Sano H, Seo S, Koizumi N, Niki T, Iwamura H, Ohashi Y (1996) Regulation of cytokinin of endogenous levels of jasmonic and salicylic acids in mechanically wounded tobacco plants. Plant Cell Physiol 37:762–769

Seo S, Okamoto M, Seto H, Ishizuka K, Sano H, Ohashi Y (1995) Tobacco MAP kinase: a possible mediator in wound signal transduction pathways. Science 270:1988–1992

Seo S, Sano H, Ohashi Y (1997) Jasmonic acid in wound signal transduction pathways. Physiol Plant 101:740–745

Seo S, Sano H, Ohashi Y (1999) Jasmonate-based wound signal transduction requires activation of WIPK, a mitogen-activated protein kinase. Plant Cell 11:289–298

Shiozaki K, Russell P (1995) Counteractive roles of protein phosphatase 2C (PP2C) and a MAP kinase kinase homolog in the osmoregulation of fission yeast. EMBO J 14:492–502

Staswick PE (1992) Jasmonate, genes, and fragrant signals. Plant Physiol 99:804–807

Stone JM, Walker JC (1995) Plant protein kinase families and signal transduction. Plant Physiol 108:451–457

Stratmann JW, Ryan CA (1997) Myeline basic protein kinase activity in tomato leaves is induced systemically by wounding and increases in response to systemin and oligosaccharide elicitors. Proc Natl Acad Sci USA 94:11085–11089

Sun H, Charles CH, Lau LF, Tonks NK (1993) MKP-1 (3CH134), an immediate early gene product, is a dual specificity phosphatase that dephosphorylates MAP kinase in vivo. Cell 75:487–493

Suzuki K, Shinshi H (1996) Protein kinases in elicitor signal transduction in plant cells. J Plant Res 109:253–263

Usami S, Banno H, Ito Y, Nishihama R, Machida Y (1995) Cutting activates a 46-kDa protein kinase in plants. Proc Natl Acad Sci USA 92:8660–8664

Wasternack C, Parthier B (1997) Jasmonate-signalled plant gene expression. Trends Plant Sci 2:302–307

Wildon DC, Thain JF, Minchin PEH, Gubb IR, Reilly AJ, Skipper YD, Doherty HM, O'Donnell PJ, Bowles DJ (1992) Electrical signalling and systemic proteinase inhibitor induction in the wounded plant. Nature 360:62–65

Wilson C, Eller N, Gartner A, Vicente O, Heberle-Bors E (1993) Isolation and characterization of a tobacco cDNA clone encoding a putative MAP kinase. Plant Mol Biol 23:543–551

Wilson C, Anglmayer R, Vicente O, Heberle-Bors E (1995) Molecular cloning. functional expression in *Escherichia coli*, and characterization of multiple mitogen-activated protein kinases from tobacco. Eur J Biochem 233:249–257

Wilson C, Voronin V, Touraev A, Vicente O, Heberle-Bors E (1997) A developmentally regulated MAP kinase activated by hydration in tobacco pollen. Plant Cell 9:2093–2100

Wurgler-Murphy SM, Maeda T, Witten EA, Saito H (1997) Regulation of the *Saccharomyces cerevisiae HOG1* mitogen-activated protein kinase by the *PTP2* and *PTP3* protein tyrosine phosphatases. Mol Cell Biol 17:1289–1297

Zhang S, Klessig DF (1997) Salicylic acid activates a 48-kDa MAP kinase in tobacco. Plant Cell 9:809–824

Zhang S, Klessig DF (1998) The tobacco wounding -activated protein kinase is encoded by *SIPK*. Proc Natl Acad Sci USA 95:7225–7230

Zhang S, Du H, Klessig DF (1998) Activation of the tobacco SIP kinase by both a cell wall-derived carbohydrate elicitor and purified proteinaceous elicitins from *Phytophthora* spp. Plant Cell 10:435–449

# Pathogen-Induced MAP Kinases in Tobacco

Shuqun Zhang[1] and Daniel F. Klessig[2]

**Summary:** The activation of two tobacco MAP kinases, SIPK and WIPK, by a variety of pathogen-associated stimuli and other stresses have been analyzed (Table 1). SIPK was activated by SA, a CWD carbohydrate elicitor and two elicitins from *Phytophthora* spp, bacterial harpin, TMV, and Avr9 from *Cladosporium fulvum*. In addition to these pathogen-associated stimuli, wounding also activated SIPK, suggesting that this enzyme is involved in multiple signal transduction pathways. In all cases tested, SIPK activation was exclusively post-translational via tyrosine and threonine/serine phosphorylation. WIPK was activated by only a subset of these stimuli, including infection by TMV or harpin-producing *Pseudomonas syringae* (preliminary unpubl. result) and treatment with the CWD elicitor, elicitins or Avr9. In contrast to SIPK, WIPK was activated at multiple levels. Low level activation (e.g. by the CWD elicitor) appeared to be primarily post-translational whereas dramatic increases in kinase activity (e.g. by TMV or elicitins) required not only post-translational phosphorylation, but also preceding rises in mRNA levels and de novo synthesis of WIPK protein. Interestingly, under conditions where the same stimulus activated both of these kinases, their kinetics of activation appeared to be distinct. SIPK was the first to be activated. Activation of the low basal level of WIPK protein present before treatment exhibited similar kinetics to that of SIPK; however, the appearance of high levels of WIPK enzyme activity was delayed, perhaps reflecting the need for WIPK transcription and de novo protein synthesis.

[1] *Present address:* 117 Schweitzer Hall, Department of Biochemistry, University of Missouri, Columbia, Missouri 65211, USA

[2] *Corresponding author:* Waksman Institute and Department of Molecular Biology and Biochemistry, Rutgers, The State University of New Jersey, 190 Frelinghuysen Road, Piscataway, New Jersey 08854-8020 USA

Results and Problems in Cell Differentiation, Vol. 27
Hirt (Ed.): MAP Kinases in Plant Signal Transduction
© Springer-Verlag Berlin Heidelberg 2000

# 1
# Introduction

Plant disease resistance and susceptibility are governed by the genotypes of both the host and the pathogen and depend on a complex exchange of signals and responses. During the long process of host-pathogen co-evolution, plants have developed various elaborate mechanisms to protect themselves against pathogen attack. Some of these defenses are preformed and provide chemical and physical barriers to inhibit pathogen infection, while others are activated only after pathogen attack. Induction of plant defense responses involves a network of signal transduction pathways and the rapid activation of gene expression following pathogen infection (Yang et al. 1997).

A critical difference between resistant and susceptible plants is the timely recognition of the invading pathogen and the rapid and effective activation of host defenses. A resistant plant is capable of rapidly activating a wide variety of defense responses that prevent pathogen colonization. In contrast, a susceptible plant exhibits much slower and weaker defense responses that fail to prevent pathogen growth and/or spread. As a result, a susceptible plant is often severely damaged or even killed by the pathogen. The activation of defense responses is initiated by host recognition of pathogen-derived molecules that are often called elicitors (e.g. microbial proteins, small peptides and oligosaccharides, etc.). Sometimes pathogen attack releases host-encoded molecules that can also serve as elicitors. There are two types of pathogen-derived elicitors – race-specific and non-race specific. A race-specific elicitor is frequently encoded by a dominant avirulence (Avr) gene. The Avr protein is recognized only by plants carrying a dominant resistance (R) gene, which may encode the receptor for the elicitor. This recognition leads to an incompatible interaction, in which the plant's defense responses are activated and the growth and/or spread of the pathogen is inhibited. This gene-for-gene interaction is receiving increasing attention with the cloning of not only Avr genes, but more recently, host R genes. Non-race-specific or non-specific elicitors induce defense responses even in plants not carrying specific R genes. Host recognition of these non-specific elicitors is likely to be mediated by high affinity, plasma membrane-associated receptors (Yang et al. 1997).

The interaction of these elicitors with host receptors probably activates a signal transduction cascade that may involve protein phosphorylation, ion fluxes, reactive oxygen species, and other signaling events (Yang et al. 1997). Many of these defense signaling pathways act through transcription factors, whose transcriptional and/or post-translational activation results in the induction of defense genes.

Defense response activation is frequently manifested at the macroscopic level as necrotic lesions. These lesions result from localized host cell death at the sites of infection and their development is termed the hypersensitive response (HR). Subsequent to HR development, the uninoculated portions of the plant frequently develop a long-lasting, broad-spectrum resistance to infection by pathogens known as systemic acquired resistance (SAR; Ryals et al. 1996; Yang et al. 1997).

The development of HR and SAR is accompanied by the activation of many plant protectant and defense genes, whose products include glutathione S-transferases, peroxidases, cell wall proteins, proteinase inhibitors, hydrolytic enzymes, and enzymes involved in phytoalexin biosynthesis, such as phenylalanine ammonia lyase (PAL) and chalcone synthase (CHS; Yang et al. 1997). The synthesis of several families of pathogenesis-related (PR) proteins also correlates well with development of both HR and SAR, particularly in tobacco and *Arabidopsis*. Furthermore, many of these proteins exhibit antimicrobial activity in vitro and/or when constitutively overexpressed in transgenic plants. In general, the acidic extracellular PRs are dramatically induced after infection, which makes them excellent markers for resistance. In addition, a large body of evidence has shown that salicylic acid (SA) is an important endogenous signal required for the activation and/or potentiation of defense responses in both local and systemic resistance (Hammond-Kosack and Jones 1996; Durner et al. 1997).

Protein kinases and phosphatases play a central function in many signal transduction pathways involved in cell differentiation, cell growth and cellular responses to environmental stimuli for both plants and animals. An increasing body of evidence indicates that phosphorylation and dephosphorylation are also involved in the activation of plant defense responses to microbial attack. For example, the development of HR in tobacco plants following tobacco mosaic virus (TMV) infection was blocked by the protein phosphatase inhibitor okadaic acid, suggesting the involvement of a dephosphorylation event (Dunigan and Madlener 1995). In contrast, fungal elicitors triggered rapid and transient protein phosphorylation in parsley and tomato suspension cells (Dietrich et al. 1990; Felix et al. 1991). This phosphorylation was blocked by the protein kinase inhibitors K-252a and staurosporine, which also suppressed elicitor-induced defense responses. Conversely, okadaic acid and two other phosphatase inhibitors, calyculin A and cantharadin, mimicked the effects of elicitors to activate certain defense responses (Felix et al. 1994; Levine et al. 1994; Mackintosh et al. 1994; Chandra and Low 1995).

Phosphorylation and/or dephosphorylation are also involved in the activation of a number of defense genes. For example, the elicitor-responsive CHS promoter of soybean contains an H-box element, which is recognized by two phosphoproteins, KAP-1 and KAP-2. Dephosphorylation of these two factors alters the mobility of their protein-DNA complexes (Yu et al. 1993). The induction of the CHS gene by bacterial infection appears to require phosphorylation of the bZIP transcription factor G/HBF-1, another H-box-binding protein. The activity of the serine/threonine kinase that phosphorylates G/HBF-1 is induced by infection and the phosphorylated factor has increased binding activity for the *chs15* promoter (Dröge-Laser et al. 1997). Similarly, elicitor-induced phosphorylation of the DNA binding factor PBF-1 enhances its binding activity for, and thereby activates, the potato *PR-10a* gene (Després et al. 1995). The elicitor-activated kinase responsible for PBF-1 phosphorylation has properties that are similar to mammalian protein kinase C (Subramaniam et al. 1997). In contrast, activation of the *PR-1* genes by SA in tobacco involves dephosphorylation of at least one phosphoprotein; this activation was blocked by okadaic acid and calyculin A (Conrath et al. 1997).

# 2
# Salicylic Acid-Induced Protein Kinase (SIPK)

Our studies with protein kinase and phosphatase inhibitors suggest that at least two phosphoproteins participate in the SA-mediated signal transduction pathway leading to *PR* gene expression – one acting upstream and the other downstream of SA (Conrath et al. 1997). These results prompted a search for an SA-induced protein kinase.

We (Zhang and Klessig 1997) discovered that in tobacco suspension cells, SA induced a very rapid and transient activation of a 48 kDa kinase, which we termed SIPK. Activity was at a maximum by 5 min and returned to basal levels by 45 min. Biologically active analogues of SA, which induce *PR* genes and enhance resistance, also activated this kinase, while inactive analogues did not. Phosphorylation of tyrosine residues of SIPK was associated with its activation. This result, together with its strong preference for myelin basic protein (MBP) as a substrate, suggested that SIPK is a member of the MAP kinase family.

To confirm or dispel this possibility, SIPK was first purified more than 25 000-fold to homogeneity from SA-treated tobacco suspension cells. The native molecular mass of the purified kinase was approximately 50 kDa, indicating that SIPK is monomeric. The purified protein was strongly phosphorylated on tyrosine residue(s). Furthermore, treatment of purified SIPK with either a tyrosine-specific phosphatase (YOP) or a serine/threonine-specific phosphatase (PP1) inactivated SIPK activity, suggesting that phosphorylation of both tyrosine and serine/threonine was required for its activation. The absolute $K_m$ and $V_{max}$ of the enzyme for $Mg^{2+}$-ATP was estimated to be 24 µM and 0.39 µmol min$^{-1}$ mg$^{-1}$, respectively. Of the proteins tested for their ability to serve as in vitro substrates, only MBP proved to be a good phosphate acceptor. Phosphoamino acid analysis by two-dimensional thin-layer electrophoresis demonstrated that only threonine residues were phosphorylated on MBP.

The sequences of six internal tryptic peptides were obtained and used to design oligonucleotide primers. After cloning a fragment of the *SIPK* gene by RT-PCR, a full-length cDNA was obtained. Analysis of the *SIPK* sequence showed that it contains all 11 conserved kinase subdomains found in serine/threonine kinases (Hanks and Hunter 1995) and it has the MAP kinase signature phosphorylation motif TXY between subdomains VII and VIII. Although SIPK is certainly a MAP kinase, it has a proline in place of an alanine in the conserved subdomain VIII. Since this alanine residue is conserved in all other MAP kinases obtained from mammals and yeast, as well as plants, SIPK is a MAP kinase distinct from other plant MAP kinases implicated in stress responses, such as WIPK.

# 3
# MAP Kinases Activated by a Cell Wall-Derived Carbohydrate Elicitor and Purified Proteinaceous Elicitins from *Phytophthora* spp.

Several lines of evidence suggest that protein phosphorylation is involved in the regulation of elicitor-induced defense responses. Treatment of plant suspension cells with fungal cell wall-derived (CWD) carbohydrate elicitors causes rapid changes in phosphoprotein profiles (Dietrich et al. 1990; Felix et al. 1991, 1994). Similar changes in phosphoprotein profiles have also been observed in response to elicitins (Viard et al. 1994), which are a family of 10 kDa proteins secreted by the pathogenic fungal genus *Phytophthora* that induce HR-like necrosis and SAR in tobacco (Yu 1995). Moreover, inhibition of phosphorylation by protein kinase inhibitors blocks elicitor-stimulated medium alkalization, generation of reactive oxygen species and activation of defense genes such as *PAL* (Grosskopf et al. 1990, Felix et al. 1991; Viard et al. 1994; Suzuki et al. 1995), while protein phosphatase inhibitors mimic the effects of the fungal elicitors (Felix et al. 1994).

## 3.1
## SIPK

Suzuki and Shinshi (1995) reported that treatment of tobacco cells with a CWD elicitor from *P. infestans* activated a 47 kDa kinase with characteristics of a MAP kinase, but the identity of this kinase and its encoding gene were unknown. Therefore, we set out to determine whether different elicitors from *Phytophthora* spp. might act through MAP kinases such as SIPK (Zhang et al. 1998). Three elicitors were tested: (1) a crude CWD carbohydrate elicitor from *P. parasitica*, (2) a purified α elicitin from *P. parasitica* termed parasiticein, and (3) a purified β elicitin from *P. cryptogea* called cryptogein.

Treatment of tobacco suspension cells with the CWD elicitor rapidly and rather transiently activated a 48 kDa kinase that preferentially used MBP as a substrate. Its activity peaked within approximately 10 min and returned to basal level between 4 and 8 h after treatment. Parasiticein and cryptogein also rapidly activated a 48 kDa kinase whose activity remained at high levels for ap-

proximately 4 h before declining in conjunction with the appearance of elicitin-induced hypersensitive cell death. The rapid activation of the 48 kDa
kinase paralleled or slightly preceded the rapid induction of medium alkalization and *PAL* gene expression induced by all three elicitors.

The size and substrate preference of the 48 kDa kinase was reminiscent of
SIPK. Its identity as SIPK was established/confirmed by three criteria. First, the
CWD elicitor-activated 48 kDa kinase co-purified through several chromatog-
raphy steps with SIPK, and second, it co-migrated with SIPK on two-dimensional gels. Third, the antibody Ab-p48N, which was raised against a peptide
corresponding to the unique N-terminus of SIPK, immunoreacted with the 48
kDa kinase activated by all three elicitors. (Note in this and all subsequent immune-complex kinase assays, the 25 µl kinase reaction contained 0.1 mg/ml
MBP and 10 micromolar ATP with 1 µCi of $[\gamma\text{-}^{32}P]$ATP.) Based on these results,
all three fungal elicitors appear to activate SIPK. However, because tobacco is
amphidiploid, it is possible that the 48 kDa MAP kinase activated by SA and
these three elicitors is encoded by both *SIPK* and its corresponding gene from
the other tobacco parent. One candidate for the corresponding gene is *Ntf4*,
which shares 93 % amino acid sequence identity with SIPK. However, we have
found by western analysis that SIPK is present in both *Nicotiana tomentosifor-
mis* and *N. sylvestris*, the parental diploid species of *N. tabacum* (unpubl. result). Moreover, Ntf4 appears to play a role in the activation of pollen grains after hydration (Wilson et al. 1997). These observations, in conjunction with the
co-purification and co-migration of SIPK and the CWD elicitor-activated 48
kDa kinase on a two-dimensional gel, argues that if there are two genes for this
48 kDa kinase, their products are either very similar or identical.

Activation of SIPK by the three elicitors and by SA was not associated with
increases in *SIPK* mRNA or protein levels. Rather, phosphorylation of a tyrosine residue(s) on SIPK correlated with its activation. Moreover, purified SIPK
could be inactivated by treatment with either a serine/threonine-specific phosphatase or a tyrosine-specific phosphatase (Zhang and Klessig 1997). Thus,
SIPK appears to be regulated exclusively at the post-translational level via threonine and tyrosine phosphorylation, just like all characterized MAP kinases
from mammals and yeast.

## 3.2
### Wounding-Induced Protein Kinase (WIPK)

Parasiticein and cryptogein treatment of tobacco also activated a 44 kDa
kinase and a 40 kDa kinase. The activation of these kinases was delayed, with
peak activity detected at approximately 4h, just before the commencement of
cell death. The 44 kDa, but not the 40 kDa, kinase was also activated by the
CWD elicitor. The extent of activation, however, was less than that seen with
the elicitins (Zhang et al. 1998).

Recently, we discovered that the 44 kDa kinase is encoded by *WIPK*. The tobacco *WIPK* (wounding induced protein kinase) gene is induced at the mRNA level by wounding (Seo et al. 1995; Zhang and Klessig 1998b). This conclusion is based on the observation that the antibody Ab-p44N, which was prepared against a peptide corresponding to the unique N terminus of WIPK, immunoreacted with the CWD elicitor- or elicitin-activated 44 kDa kinase in an immune-complex kinase assay (unpubl. result). The identity of the 40 kDa kinase is currently unclear.

In contrast to SIPK, elicitin-induced increases in kinase activity of WIPK were preceded by dramatic increases in *WIPK* mRNA and protein levels. This differs from the weak activation of WIPK induced by the CWD elicitor, where a low level of enzyme activation preceded increases in *WIPK* mRNA and protein. Surprisingly, the CWD elicitor-mediated accumulation of *WIPK* mRNA and protein did not result in a further increase in kinase activity. Thus, activation of WIPK appears to be regulated at multiple steps.

Treatment of tobacco cells with either the elicitins or the CWD elicitor induced medium alkalization and *PAL* gene expression, two rapid responses associated with disease resistance. After considerable delay, the elicitins also induced cell death. The rapid kinetics of SIPK activation are consistent with its involvement in the two early responses, while the slower activation kinetics of WIPK parallel the delayed cell death induced by the elicitins. In addition, prevention of SIPK and WIPK activation with staurosporine suppressed CWD elicitor-induced medium alkalization and *PAL* expression (Zhang et al. 1998) and elicitin-stimulated cell death (unpubl. result). Taken together, these kinetic and inhibitor studies suggest SIPK's participation in the two early responses and WIPK's involvement in cell death.

The fungal elicitors are unlikely to activate SIPK through increases in SA levels because SIPK activation by these elicitors was very rapid, with peak activity reached within 5 to 10 min. These initial kinetics of activation were very similar to those induced by exogenous SA. If SIPK activation by these elicitors was mediated by SA, activation would probably be delayed due to the need for SA synthesis. Activation of WIPK by these three elicitors also does not appear to involve SA as SA treatment itself does not activate WIPK at the mRNA, protein or enzyme levels (Zhang and Klessig 1997; unpubl. results).

What is the relationship between the 47 kDa kinase activated by the CWD elicitor from *P. infestans* (Suzuki and Shinshi 1995) and the 48 kDa SIPK and 44 kDa WIPK activated by the CWD elicitor from *P. parasitica* (Zhang et al. 1998; unpubl. result)? Based on the similarities in size and activation kinetics, we suspect that the 47 kDa kinase corresponds to SIPK. Moreover, this is consistent with the fact that SIPK is the predominant kinase activity induced by the *P. parasitica* CWD elicitor and detected via an in-gel kinase activity assay with MBP as the substrate.

# 4
# Activation of SIPK by Bacterial Harpin

Harpins are bacterial encoded protein elicitors that induce HR-like necrosis on plants such as tobacco which do not serve as a host for the bacterium (Wei et al. 1992). They have been isolated and characterized from *Erwinia amylovora* (Wei et al. 1992) and *Pseudomonas syringae* pv *syringae* (He et al. 1993). Novacky and colleagues (Ádám et al. 1997) found that infiltration of harpin purified from *E. amylovora* into tobacco leaves activated a 48 kDa kinase which they termed HAPK (for harpin-activated protein kinase). Activation was rapid, with activity peaking between 10 and 15 min, and transient. HAPK's preference for MBP as a substrate and requirement for tyrosine phosphorylation for activity suggested that it was a MAP kinase. Similarly, we observed that infection of tobacco leaves with *P. syringae* pv *syringae* strain 61 activated a 48 kDa kinase (unpubl. result). Infection with the *Pss61 hrpH⁻* mutant failed to activate this kinase, suggesting that activation required the harpin elicitor. In a collaborative effort, HAPK was recently identified as SIPK using the SIPK-specific antibody Ab-p48N in an immune-complex kinase assay (Hoyos et al. 1998).

# 5
# Activation of SIPK and WIPK by TMV

The *Phytophthora* CWD elicitor and elicitins, as well as the bacterial harpins, are non-race-specific elicitors that induce a variety of defense responses. These include medium alkalization, *PAL* gene expression, and necrosis on plants which do not normally serve as a host for the pathogen (Fig. 1). In addition to this type of non-host resistance, which does not require specific resistance genes, some plant-pathogen interactions are mediated by a so-called gene-for-gene recognition which was first proposed by Flor (1971). In this scenario, host perception of the invading pathogen is mediated by the product of a dominant *R* gene in the host and its corresponding dominant *Avr* gene in the pathogen. For example, the tomato *Cf-9* gene mediates resistance only towards those races of *Cladosporium fulvum* that express the *Avr9* gene.

During the past 6 years, many *R* genes have been cloned. Interestingly, although these *R* gene products confer specific resistance to different viral, bacterial, fungal or nematode pathogens, most contain one or more common structural motifs (Bent 1996; Baker et al. 1997; De Wit 1997). The most common features of R proteins are the leucine-rich repeats (LRR). LRR have been implicated in protein-protein interactions and may be responsible for recognition of the corresponding Avr factor (race-specific elicitor) in those cases in which the *R* gene encodes the receptor for the corresponding Avr factor.

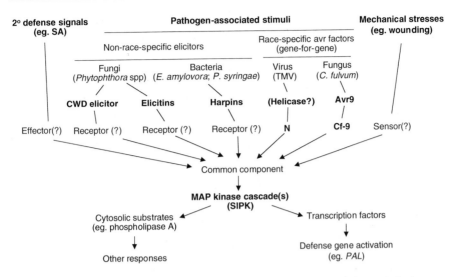

**Fig. 1.** One possible model for SIPK activation. We suggest that a variety of stimuli feed into a common component that initiates the SIPK cascade. These stimuli include the secondary defense signal SA and various pathogen-associated stimuli such as several non-race-specific bacterial or fungal elicitors and direct or indirect gene-for-gene interactions between resistance and avirulence gene products. Wounding, which is often associated with insect predation, also activates SIPK. The activated SIPK presumably induces different cellular responses such as defense gene activation through phosphorylation of different substrates which may include nuclear transcription factors or cytosolic substrates such as phospholipase A. Different cellular responses induced by different stimuli may be determined by the duration and/or magnitude of SIPK activation and/or the parallel activation of other kinases such as WIPK

In tobacco, resistance to TMV is conditioned by the resistance gene *N*, which has been recently cloned. TMV infection of tobacco cultivars carrying the *N* gene, such as Xanthi nc (NN), causes a large increase in endogenous SA levels, activation of a complex array of *PR* genes and development of an HR (Malamy and Klessig 1992). These defense responses, along with resistance to TMV, can be reversibly blocked at high temperatures (e.g. 32 °C). Upon shifting plants to lower temperatures (e.g. 22 °C), all of these defense responses are rapidly and strongly induced in a more synchronous manner than occurs when the entire infection procedure is carried out at 22 °C (Malamy et al. 1992). We have taken advantage of this reversible, high temperature inhibition of TMV-induced defense responses to follow changes in kinase activity (Zhang and Klessig 1998a). Within 4 h after shifting infected tobacco plants from 32 °C to 22 °C, a 48 kDa kinase and a 44 kDa kinase were activated in TMV-, but not in mock-infected plants. The size and substrate preference of these two kinases suggested that they might be SIPK and WIPK, respectively. This was confirmed using the SIPK-specific antibody Ab-p48N and the WIPK-specific antibody Ab-p44N in immune-complex kinase assays.

Activation of both SIPK and WIPK by TMV required phosphorylation of tyrosine and serine/threonine, arguing that both enzymes are regulated post-translationally. As with SA, harpin and the fungal elicitors, activation of SIPK by TMV was exclusively post-translational. No changes in *SIPK* mRNA or protein levels were evident. In contrast, increases in WIPK enzyme activity were preceded by large rises in both mRNA and protein levels. Thus, WIPK's activation by TMV (as well as by elicitins) requires multiple steps including de novo synthesis of its protein.

WIPK activation by TMV was dependent on the *N* resistance gene, as infection of the susceptible Xanthi (nn) cultivar failed to activate WIPK. Furthermore, *WIPK* mRNA levels increased not only in the TMV-inoculated leaves but also systemically in uninoculated leaves of the infected plant. This result suggests that WIPK may play a role in the establishment of SAR.

Although many *N* gene-mediated defense responses are SA dependent, activation of WIPK is SA independent. Not only did SA treatment fail to activate WIPK, its activation at the mRNA, protein and enzyme levels following TMV infection was essentially identical in wild type Xanthi nc (NN) plants and transgenic Xanthi nc (NN) plants, which fail to accumulate SA due to expression of the *NahG* transgene. Whether WIPK is a component upstream of SA in an *N* gene-mediated defense signaling pathway, or whether it participates in an SA-independent pathway is presently unclear.

# 6
# Activation of SIPK and WIPK by Avr9
# from *Cladosporium fulvum*

Jones and co-workers (Hammond-Kosack et al. 1998) have transferred the *Cf-9* gene of tomato into tobacco and potato. Treatment of these transgenic plants with Avr9 induced an HR, arguing that essential components for Cf-9-mediated signal transduction are conserved among these three Solanaceae.

To elucidate the role of kinases in Cf-9-mediated defense responses induced by Avr9, changes in kinase activity were monitored with an in-gel kinase assay (Romeis et al. 1999). Leaves from Cf-9 transgenic tobacco were injected with either intercellular fluid (IF) from transgenic tobacco leaves expressing the Avr9 peptide apoplastically or with a control IF without Avr9. Treatment with the Avr9[+] IF resulted in a rapid and transient activation of a 48 kDa kinase and a 46 kDa kinase which peaked within 15 min and returned to near basal levels by 2–3 h. The activity of the 48 kDa kinase was much stronger than that of the 46 kDa kinase. Injection of Avr9[−] IF caused a weaker and even more transient activation of the 48 kDa kinase, which appears to be due to the osmotic and/or mechanical stresses associated with injection. A similar activation of this kinase was previously observed in water-injected tobacco leaves (see Sect. 7; Zhang et al. 1998). Treatment of suspension cells derived from the Cf-9 transgenic tobacco with Avr9[+] IF also resulted in a rapid and transient activation of these two kinases, and their activity levels were more similar than those seen in leaves. No activation was detected with the Avr9[-] IF or when Avr9[+] IF was added to suspension cells of nontransgenic tobacco. To confirm that the Avr9 peptide in the IF was responsible for activation, synthetic Avr9 was shown to activate both kinases in suspension cells and leaves of Cf-9 tobacco. Thus, two kinases were transiently and rapidly activated in an Avr9/Cf-9-specific manner in accordance with the gene-for-gene hypothesis.

The size and preference for MBP as a substrate suggested that the Avr9-activated 48 kDa kinase and 46 kDa kinase might be SIPK and WIPK, respectively. This was verified by using a combination of immunoprecipitation with the SIPK-specific antibody Ab-p48N and the WIPK-specific antibody Ab-p44N, followed by an in-gel kinase assay.

Activation of both SIPK and WIPK in Cf-9 tobacco by Avr9 was associated with phosphorylation of at least one of their tyrosine residues. Interestingly, tyrosine phosphorylation remained high even 2–3 h after Avr9 treatment, although the activity of both kinases had returned to basal levels. This result suggests that inactivation under these conditions may be achieved by other mechanisms which do not involve dephosphorylation of tyrosine as the first regulatory step.

Activation of SIPK by Avr9 was solely post-translational, just as was previously observed in response to SA, TMV and the non-specific elicitors. No changes in SIPK protein levels were detected. Similarly, WIPK protein levels remained unchanged after activation. However, by 30 min after Avr9 treatment, *WIPK* mRNA levels began to increase. Surprisingly, this elevated level of *WIPK* mRNA did not result in a higher amount of WIPK protein, nor was it responsible for elevated WIPK enzyme activity since activation was evident by 2–5 min after treatment and peaked by 15 min.

The kinetics of WIPK activation by Avr9 appear to differ from those induced by TMV. With TMV, a large increase in WIPK activity was preceded by SIPK activation and large increases in both *WIPK* mRNA and protein levels. In contrast, following Avr9 treatment, WIPK and SIPK exhibited similar kinetics of activation and the increase in WIPK activity was not dependent on changes in *WIPK* mRNA or protein levels. This is somewhat reminiscent of the CWD elicitor-mediated activation of WIPK, which was transient and preceded changes in mRNA. However, with the CWD elicitor (unpubl. result), both WIPK protein and mRNA levels increased and the amount of WIPK activity appeared to be much lower than that induced by Avr9 in suspension cells. One possible explanation for some of these differences is the difference in basal levels of WIPK protein in the Xanthi (nc) nontransgenic tobacco versus the Havana petite cultivar transformed with the *Cf-9* gene. In the nontransgenic Xanthi suspension cells, the basal levels of SIPK to WIPK protein were estimated at approximately 10:1 (Zhang and Klessig 1998b) while in the Cf-9 transgenic suspension cells approximately equal levels of these two MAP kinases were found (Romeis et al.

1999). Some of the differences noted above may also reflect the use of a viral pathogen versus treatment with a fungal-derived factor.

The similarities between defense responses activated by different *R* gene products, as well as the conservation of structural motifs observed in these proteins, suggest that diverse plant-pathogen interactions activate a common signal transduction pathway(s) leading to disease resistance. Given that SIPK and WIPK are activated in wild type tobacco resisting TMV infection or Cf-9 transgenic tobacco responding to Avr9, it seems likely that their orthologs may participate in the defense response pathway(s) initiated by *R* gene products found in other plant species. In addition, defense signals from *R* genes within a plant species, as well as non-race-specific elicitors such as elicitins and harpins, may converge on the same MAP kinase cascade(s). Thus, determination of the role(s) of these MAP kinases in *R* gene-mediated defense signaling should enhance our understanding of the pathway(s) leading to resistance and may provide new opportunities for enhancing disease resistance in plants.

# 7
# Wounding and MAP Kinase Activation

Several kinase activities which are induced by wounding have been reported. They are believed to be MAP kinases because they are themselves phosphorylated on tyrosine residues upon activation and they preferentially use MBP as a substrate. These include the 46 kDa wounding-activated (Seo et al. 1995) or cutting-activated (Usami et al. 1995) kinase from tobacco and the wounding-, systemin-, and oligosaccharide- activated 48 kDa kinase from tomato (Stratmann and Ryan 1997). As noted earlier, *WIPK*, which was isolated from a TMV-infected tobacco cDNA library, was shown to be rapidly induced at the mRNA level by wounding (Seo et al. 1995). This result led to the hypothesis that *WIPK* encoded the 46 kDa wounding-activated kinase.

Experiments designed to determine whether SA or the CWD fungal elicitor activated SIPK in whole plants as well as in tobacco suspension cells revealed the presence of a wounding-inducible 48 kDa kinase (Zhang and Klessig 1998b). When either SA or the CWD fungal elicitor were infiltrated into tobacco leaves, a 48 kDa kinase, which was recognized by the SIPK-specific antibody Ab-p48N, was activated. Unexpectedly, infiltration of the water control also activated a 48 kDa kinase, but with different kinetics from those of the SA-induced or CWD elicitor-induced kinase. To test our suspicion that wounding caused by infiltration was responsible for the water-induced activation of the 48 kDa kinase, two other wounding methods were tried. Both wounding by rubbing with carborundum á la Seo et al. (1995) or cutting á la Usami et al. (1995) resulted in the rapid and transient activation of a 48 kDa kinase. Activation peaked at approximately 5 min and activity returned to the basal level in less than 1 h.

To determine if this 48 kDa wounding-activated kinase was encoded by *SIPK* or by *WIPK*, the SIPK-specific antibody Ab-p48N and the WIPK-specific antibody Ab-p44N were employed in immune-complex kinase assays (using 10 micromolar ATP in the kinase reaction). For all three treatments (water infiltration, abrading and cutting), the 48 kDa kinase was immunoprecipitated by Ab-p48N, but not Ab-p44N, arguing that this kinase is encoded by *SIPK*. As with other stimuli, wounding-induced activation of SIPK was exclusively post-translational via phosphorylation. Interestingly, all three treatments led to a transient increase in *WIPK* mRNA levels. However, little or no increase in WIPK protein or enzyme activity were evident. This increase in *WIPK* mRNA was first detected within 20–30 min of treatment, which was significantly later than the wounding-induced activation of the 48 kDa SIPK.

Our initial studies with pathogens, pathogen-derived elicitors and the plant defense secondary signal SA argued that SIPK participates in the induction of disease resistance responses. SIPK's activation by wounding suggests that this same MAP kinase is involved in a second pathway leading to wounding responses.

**Table 1.** Activation of SIPK and WIPK at multiple levels by various stimuli

| Stimuli | SIPK | | | WIPK | | |
|---|---|---|---|---|---|---|
| | mRNA | Protein ↑ | Activity ↑[a] | mRNA | Protein ↑ | Activity ↑[a] |
| SA[h,i] | −[f] | − | ++ | − | − | − |
| CWD elicitor[b,i] | − | − | ++ | ++ | ++ | −/+[g] |
| Elicitins[c,h,i] | − | − | ++ | ++ | ++ | ++ |
| Harpin[d,h] | − | − | ++ | NT | NT | NT |
| TMV[h] | − | − | ++ | ++ | ++ | ++ |
| Avr9[e,h,i] | NT | − | ++ | ++ | − | ++/+ |
| Wounding | − | − | ++ | + | −/+ | −/+[g] |

The data above are taken exclusively from studies by the authors except where collaborations are noted (i.e. harpin and *Avr9*).

[a] Kinase activity was measured by immunoselecting WIPK or SIPK with the WIPK-specific antibody Ab-p44N or the SIPK-specific antibody Ab-p48N and then performing a kinase assay with MBP 0.1 μg/ml and 10 micromolar ATP with 1 μCi of [γ-$^{32}$P] ATP as substrates.

[b] The cell wall-derived (CWD) carbohydrate elicitor was from *Phytophthora parasitica*.

[c] Purified elicitins were from *P. parasitica* (parasiticein) and *P. cryptogea* (cryptogein).

[d] Purified harpin from *Erwinia amylovora* (Hoyos et al. 1998); or infection with *Pseudomonas syringae* pv springae strain 61 with wild type harpin but not with strain Pss61 hrpH⁻.

[e] Recipient was transgenic tobacco expressing the *Cf-9* gene from tomato that confers resistance to *Cladosporium fulvum* expressing *Avr9* (Romeis et al. 1999).

[f] ++ = high level activation; + = low level activation; ± = little or no activation; − = no activation; NT = not tested.

[g] Weak, transient activation which does not correlate with increases in *WIPK* mRNA and protein.

[h] Tested in planta.

[i] Tested in suspension cell culture.

# 8
# General Discussion

SIPK's function(s) after wounding (mechanically or by insect predation) or microbial pathogen attack is presently unclear. As in mammalian and yeast systems, activated SIPK might be translocated to the nucleus where it could phosphorylate transcription factors leading to the activation of one or more defense genes. Alternatively, it might phosphorylate and thus activate cytosolic proteins. Some mammalian MAP kinases, for example, activate cytoplasmic phospholipase $A_2$ in response to injury and pathogen attack (Lin et al. 1993). The activated phospholipase $A_2$ cleaves phospholipids in the membrane, thereby releasing arachidonic acid. Arachidonic acid is the precursor to prostaglandins and leucotrienes, which are important signals involved in activating inflammatory responses. Interestingly, plants have an analogous oxylipin pathway which is activated by wounding and fungal elicitors. Activation of this pathway causes the major plant membrane fatty acid linolenic acid to be released and converted to jasmonic acid. Jasmonic acid, its methyl ester and several intermediates in this octadecanoid pathway are important secondary signals involved in activating certain defense responses (Doares et al. 1995; Bergey et al. 1996; Creelman and Mullet 1997; Wasternack and Parthier 1997). In soybean cells, harpin and a CWD elicitor from the pathogenic fungus *Verticillium dahliae* have been shown to induce a cytosolic phospholipase A activity (Chandra et al. 1996). Perhaps the soybean ortholog of SIPK activates this phospholipase A in response to these elicitors.

At present, we do not know if other stresses besides wounding and pathogen-associated stimuli activate SIPK. Perhaps SIPK is a functional homolog of the mammalian stress-activated protein kinase/Jun N-terminal kinase (SAPK/JNK) and/or the p38 kinase. These proteins belong to two distinct subfamilies of mammalian MAP kinases, and they are activated by an extensive list of stimuli including inflammatory cytokines, DNA damage, UV-C, oxidant stress, heat stress, hyperosmolarity, mechanical stretching and translational inhibition (Komuro et al. 1996; Kyriakis and Avruch 1996a). Both SAPK/JNK and p38 kinase have unique dual-phosphorylation activation motifs (TPY and TGY, respectively), compared with the classical MAP kinases involved in cell proliferation, which contain a TEY phosphorylation motif and are represented by ERK1/2 (Seger and Krebs 1995). The sequence of this activation motif is the hallmark for the classification of the three subfamilies of the mammalian MAP kinase. This sequence has also been shown to be important for their function (Jiang et al. 1997). In plants, all cloned MAP kinases so far reported have the TEY motif, including SIPK and WIPK (Seo et al. 1995; Zhang and Klessig 1997).

How does a single MAP kinase responding to diverse stimuli activate different intracellular responses? Studies in mammalian systems suggest that the duration and/or magnitude of MAP kinase activation is at least partially responsible for signaling different responses through the same kinase (Marshall 1995;

Chen et al. 1996). An excellent example is the response of PC12 cells to treatment with two different hormones; the responses to both hormones are mediated through ERK. Nerve growth factor causes prolonged ERK activation (several hours) which leads to cell differentiation (outgrowth of neurites) and cessation of cell growth. In contrast, treatment of PC12 cells with epidermal growth factor results in a transient activation of ERK which stimulates cell proliferation (Marshall 1995). Similarly, rapid and transient activation of SAPK/JNK kinase by different stress stimuli leads to responses which allow cells to adapt to their environment (Kyriakis and Avruch 1996a), while sustained activation leads to apoptosis (Chen et al. 1996).

Activation of SIPK by SA and the three fungal elicitors was most readily compared in tobacco suspension cells. While the magnitude of SIPK activation was similar for all four treatments, the kinetics were very different. SA induced a rapid and transient activation, with activity peaking by 5 min and returning to basal level within 45 min (Zhang and Klessig 1997). The CWD elicitor caused a rapid but more prolonged activation, which remained at half maximal up to 2 h after treatment. In contrast, activation of SIPK by the two elicitins persisted at high levels until cell death (Zhang et al. 1998).

Alternatively or in addition, differential activation of the 44 kDa WIPK and unidentified 40 kDa kinase by these four treatments could be important in specifying the cellular responses (Zhang et al. 1998). While SA activated only the 48 kDa SIPK, the CWD elicitor activated SIPK and WIPK, and the two elicitins stimulated all three kinases. Furthermore, CWD elicitor activation of WIPK was more transient and weaker than activation by the elicitins and it did not correlate with the delayed increase in WIPK protein. Thus, the CWD elicitor appeared to activate only the constitutive, basal level of WIPK protein present before treatment.

As in yeast and mammals, MAP kinases in plants are encoded by multigene families (Hirt 1997; Mizoguchi et al. 1997). Analysis of a phylogenetic tree (Fig. 2) based on sequence homology among all of the cloned plant MAP kinases indicates that there are several distinct groups or subfamilies. The distribution of different family members from a particular species among the different groups argues that each group has a different function(s) and that kinases within a group share a similar function(s). One group, many of whose members have been shown to respond to different stresses, is formed by WIPK and its orthologs from alfalfa (MMK4; Jonak et al. 1996), Arabidopsis (AtMPK3; Mizoguchi et al. 1996) and parsley (ERMK; Ligterink et al. 1997). In contrast to yeast and mammalian MAP kinases, all four of these plant MAP kinase genes are induced transiently at the mRNA level by various stresses (Seo et al. 1995; Jonak et al. 1996; Mizoguchi et al. 1996; Bögre et al. 1997; Ligterink et al. 1997; Zhang and Klessig 1998a). AsMAP1 of oats also fits into this group; its mRNA level is suppressed by gibberellin treatment of aleurone cells (Huttly and Phillips 1995). However, changes in protein levels have been demonstrated only for WIPK (Zhang and Klessig 1998a).

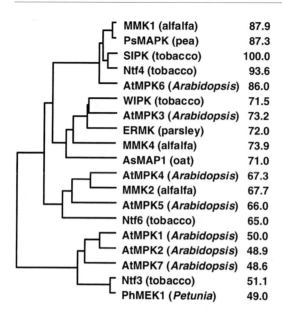

| | |
|---|---|
| MMK1 (alfalfa) | 87.9 |
| PsMAPK (pea) | 87.3 |
| SIPK (tobacco) | 100.0 |
| Ntf4 (tobacco) | 93.6 |
| AtMPK6 (*Arabidopsis*) | 86.0 |
| WIPK (tobacco) | 71.5 |
| AtMPK3 (*Arabidopsis*) | 73.2 |
| ERMK (parsley) | 72.0 |
| MMK4 (alfalfa) | 73.9 |
| AsMAP1 (oat) | 71.0 |
| AtMPK4 (*Arabidopsis*) | 67.3 |
| MMK2 (alfalfa) | 67.7 |
| AtMPK5 (*Arabidopsis*) | 66.0 |
| Ntf6 (tobacco) | 65.0 |
| AtMPK1 (*Arabidopsis*) | 50.0 |
| AtMPK2 (*Arabidopsis*) | 48.9 |
| AtMPK7 (*Arabidopsis*) | 48.6 |
| Ntf3 (tobacco) | 51.1 |
| PhMEK1 (*Petunia*) | 49.0 |

**Fig. 2.** Relationship among all cloned plant MAP kinase members. The phylogenetic tree shown as a dendrogram was created by the clustal method (MegaAlign program; DNAStar, Madison, WI). The deduced amino acid sequences are from cDNA clones of *MMK1* (also termed *MsERK1* or *Msk7*; Duerr et al. 1993; Jonak et al. 1993), *AtMPK1* to *AtMPK7* (Mizoguchi et al. 1993), *PsMAPK* (also called D5; Stafstrom et al. 1993; Pöpping et al. 1996), *Ntf3* (Wilson et al. 1993), *As-MAP1* (Huttly and Phillips 1995), *PMEK1* (Decroocq-Ferrant et al. 1995), *MMK2* (Jonak et al. 1995), *WIPK* (Seo et al. 1995), *Ntf4* and *Ntf6* (Wilson et al. 1995), *MMK4* (Jonak et al. 1996), *ERMK* (Ligterink et al. 1997) and *SIPK* (Zhang and Klessig 1997). The numbers at right indicate the percentage identity at the amino acid level with SIPK

The second group of plant MAP kinases, some of whose members are stress activated, is represented by *SIPK* (tobacco), *Ntf4* (tobacco, Wilson et al. 1995, 1997), *MMK1* (alfalfa, Duerr et al. 1993; Jonak et al. 1993), *AtMPK6* (*Arabidopsis*, Mizoguchi et al. 1993) and *PsMAPK* (pea, Stafstrom et al. 1993; Pöpping et al. 1996). In addition to similarity in sequence, members of this group are larger compared to their counterparts in the first group; this is due to a short extension at the N-terminus of the protein.

Originally, it was proposed that *WIPK* and its ortholog from alfalfa, *MMK4*, encode the wounding-activated MAP kinases (Seo et al. 1995; Bögre et al. 1997). In light of the recent demonstration that *SIPK*, rather than *WIPK*, encodes the tobacco wounding-activated kinase (Zhang and Klessig 1998b), the hypothesis that the wounding-activated kinase in alfalfa is encoded by *MMK4* needs to be reconsidered. This hypothesis was based on the demonstration that the alfalfa wounding-activated kinase was recognized by the M7 antibody, which was raised against a 10-amino acid peptide corresponding to the C-terminus of MMK4 (Jonak et al. 1996; Bögre et al. 1997). This M7 antibody recognizes recombinant MMK4 but not MMK2 or MMK3. Unfortunately, it was not tested against MMK1, which shares eight out of the ten C-terminal amino acids with MMK4. This antibody also recognized a recombinant protein encoded by *ERMK*, a *WIPK* ortholog from parsley (Ligterink et al. 1997; *ERMK* is thought to encode the *Phytophthora sojae* Pep25 elicitor-activated kinase since this kinase, like the recombinant *ERMK*-encoded protein, is recognized by M7). However, since ERMK shares only seven out of the ten C-terminal amino acids with alfalfa MMK4, while MMK1 shares eight out of the ten, it seems likely that the M7 antibody reacts with members of both the WIPK and SIPK groups or subfamily. Given that in tobacco the major *Phytophthora* elicitor-activated kinase is encoded by *SIPK* (Zhang et al. 1998), as is the wounding-activated kinase, the likelihood that *SIPK* subfamily members rather than the *WIPK* subfamily members encode these kinases should be tested.

*Acknowledgment.* The work by the authors described in this chapter was supported by a National Science Foundation Grant No. MCB9310371/MCB9723952 and United States Department of Agriculture Grant No. 9 802 200 to D.F.K.

# References

Ádám AL, Pike S, Hoyos ME, Stone JM, Walker JC, Novacky A (1997) Rapid and transient activation of a myelin basic protein kinase in tobacco leaves treated with harpin from *Erwinia amylovora*. Plant Physiol 115:853–861

Baker B, Zambryski P, Staskawicz B, Dinesh-Kumar SP (1997) Signaling in plant-microbe interactions. Science 276:726–733

Bent AF (1996) Plant disease resistance genes: Function meets structure. Plant Cell 8:1757–1771

Bergey DR, Howe GA, Ryan CA (1996) Polypeptide signaling for plant defensive genes exhibits analogies to defense signaling in animals. Proc Natl Acad Sci USA 93:12053–12058

Bögre L, Ligterink W, Meskiene I, Barker PJ, Heberle-Bors E, Huskisson NS, Hirt H (1997) Wounding induces the rapid and transient activation of a specific MAP kinase pathway. Plant Cell 9:75–83

Chandra S, Low PS (1995) Role of phosphorylation in elicitation of the oxidative burst in cultured soybean cells. Proc Natl Acad Sci USA 92:4120–4123

Chandra S, Heinstein PF, Low PS (1996) Activation of phospholipase A by plant defense elicitors. Plant Physiol 110:979–986

Chen Y-R, Meyer CF, Tan T-H (1996) Persistent activation of c-Jun N-terminal kinase 1 (JNK1) in γ radiation-induced apoptosis. J Biol Chem 271:631–634

Conrath U, Silva H, Klessig DF (1997) Protein dephosphorylation mediates salicylic acid-induced expression of *PR-1* genes in tobacco. Plant J 11:747–757

Creelman RA, Mullet JE (1997) Oligosaccharins, brassinolides, and jasmontes: nontraditional regulators of plant growth, development, and gene expression. Plant Cell 9:1211–1223

Decroocq-Ferrant V, Decroocq S, Van Went J, Schmidt E, Kreis M (1995) A homologue of the MAP/ERK family of protein kinase genes is expressed in vegetative and in female reproductive organs of *Petunia hybrida*. Plant Mol Biol 27:339–350

Després C, Subramaniam R, Matton DP, Brisson N (1995) The activation of the potato *PR-10a* gene requires the phosphorylation of the nuclear factor PBF-1. Plant Cell 7:589–598

De Wit PJGM (1997) Pathogen avirulence and plant resistance: a key role for recognition. Trends Plant Sci 2:452–458

Dietrich A, Mayer JE, Hahlbrock K (1990) Fungal elicitor triggers rapid, transient, and specific protein phosphorylation in parsley cell suspension cultures. J Biol Chem 265:6360–6368

Doares SH, Syrovets T, Weiler EW, Ryan CA (1995) Oligogalacturonides and chitosan activate plant defensive genes through the octadecanoid pathway. Proc Natl Acad Sci USA 92: 4095–4098

Dröge-Laser W, Kaiser A, Lindsay WP, Halkier BA, Loake GJ, Doerner P, Dixon RA, Lamb C (1997) Rapid stimulation of a soybean protein-serine kinase which phosphorylates a novel bZIP DNA-binding protein, G/HBF-1, during the induction of early transcription-dependent defenses. EMBO J 16:726–738

Duerr B, Gawienowski M, Ropp T, Jacobs T (1993) MsERK1: a mitogen-activated protein kinase from a flowering plant. Plant Cell 5:87–96

Dunigan DD, Madlener JC (1995) Serine/threonine protein phosphatase is required for tobacco mosaic virus-mediated programmed cell death. Virology 207:460–466

Durner J, Shah J, Klessig DF (1997) Salicylic acid and disease resistance in plants. Trends Plant Sci 2:266–274

Felix G, Grosskopf DG, Regenass M, Boller T (1991) Rapid changes of protein phosphorylation are involved in transduction of the elicitor signal in plant cells. Proc Natl Acad Sci USA 88: 8831–8834

Felix G, Regenass M, Spanu P, Boller T (1994) The protein phosphatase inhibitor calyculin A mimics elicitor action in plant cells and induces rapid hyperphosphorylation of specific proteins as revealed by pulse labeling with [$^{33}$P] phosphate. Proc Natl Acad Sci USA 91:952–956

Flor HH (1971) Current status of gene-for-gene concept. Annu Rev Phytopathol 9:275–296

Grosskopf DG, Felix G, Boller T (1990) K-252a inhibits the response of tomato cells to fungal elicitors *in vivo* and their microsomal protein kinase *in vitro*. FEBS Lett 275:177–180

Hammond-Kosack KE, Jones JDG (1996) Resistance gene-dependent plant defense responses. Plant Cell 8:1773–1791

Hammond-Kosack KE, Tang S, Harrison K, Jones JDG (1998) The tomato Cf-9 disease resistance gene functions in tobacco and potato to confer responsiveness to the fungal avirulence gene product Avr9. Plant Cell 10:1251–1266

Hanks SK, Hunter T (1995) The eukaryotic protein kinase superfamily: kinase (catalytic) domain structure and classification. FASEB J 9:576–596

He SY, Huang HC, Collmer A (1993) *Pseudomonas syringae* pv *syringae* harpin$_{Pss}$ a protein that is secreted via the *hrp* pathway and elicits the hypersensitive response in plants. Cell 73: 1255–1266

Hirt H (1997) Multiple roles of MAP kinases in plant signal transduction. Trends Plant Sci 2: 11–15

Hoyos ME, Zhang S, Johal GS, Klessig DF, Hirt H, Pike SM, Novacky AJ (1998) Is harpin-activated protein kinase related to salicylic acid- or fungal elicitor-induced kinases? Plant Physiol Suppl pp 46–47

Huttly AK, Phillips AL (1995) Gibberellin-regulated expression in oat aleurone cells of two kinases that show homology to MAP kinase and a ribosomal protein kinase. Plant Mol Biol 27: 1043–1052

Jiang Y, Li Z, Schwarz EM, Lin A, Guan K, Ulevitch RJ, Han J (1997) Structure-function studies of p38 mitogen-activated protein kinase. Loop 12 influences substrate specificity and autophosphorylation, but not upstream kinase selection. J Biol Chem 272:11096–11102

Jonak C, Páy A, Bögre L, Hirt H, Heberle-Bors E (1993) The plant homologue of MAP kinase is expressed in a cell cycle-dependent and organ-specific manner. Plant J 3:611–617

Jonak C, Kiegerl S, Lloyd C, Chan J, Hirt H (1995) MMK2, a novel alfalfa MAP kinase, specifically complements the yeast MPK1 function. Mol Gen Genet 248:686–694

Jonak C, Kiegerl S, Ligterink W, Barker PJ, Huskisson NS, Hirt H (1996) Stress signaling in plants: a mitogen-activated protein kinase pathway is activated by cold and drought. Proc Natl Acad Sci USA 93:11274–11279

Komuro I, Kudo S, Yamazaki T, Zou Y, Shiojima I, Yazaki Y (1996) Mechanical stretch activates the stress-activated protein in cardiac myocytes. FASEB J 10:631–636

Kyriakis JM, Avruch J (1996a) Protein kinase cascades activated by stress and inflammatory cytokines. Bioessays 18:567–577

Kyriakis JM, Avruch J (1996b) Sounding the alarm: protein kinase cascades activated by stress and inflammation. J Biol Chem 271:24313–24316

Levine A, Tenhaken R, Dixon R, Lamb C (1994) $H_2O_2$ from the oxidative burst orchestrates the plant hypersensitive disease resistance response. Cell 79:583–595

Ligterink W, Kroj T, zur Nieden U, Hirt H, Scheel D (1997) Receptor-mediated activation of a MAP kinase in pathogen defense of plants. Science 276:2054–2057

Lin L-L, Wartmann M, Lin AY, Knopf JL, Seth A, Davis RJ (1993) cPLA$_2$ is phosphorylated and activated by MAP kinase. Cell 72:269–278

Mackintosh C, Lyon GD, Mackintosh RW (1994) Protein phosphatase inhibitors activate antifungal defense responses of soybean cotyledons and cell cultures. Plant J 5:137–147

Malamy J, Klessig DF (1992) Salicylic acid and plant disease resistance. Plant J 2:643–654

Malamy J, Hennig J, Klessig DF (1992) Temperature-dependent induction of salicylic acid and its conjugates during the resistance response to tobacco mosaic virus infection. Plant Cell 4:359–366

Marshall CJ (1995) Specificity of receptor tyrosine kinase signaling: transient versus sustained extracellular signal-regulated kinase activation. Cell 80:179–185

Mizoguchi T, Hayashida N, Yamaguchi-Shinozaki K, Kamada H, Shinozaki K (1993) ATMPKs: a gene family of plant MAP kinases in *Arabidopsis thaliana*. FEBS Lett 336:440–444

Mizoguchi T, Irie K, Hirayama T, Hayashida N, Yamaguchi-Shinozaki K, Matsumota K, Shinozaki K (1996) A gene encoding a mitogen-activated protein kinase kinase kinase is induced simultaneously with genes for a mitogen-activated protein kinase and an S6 ribosomal protein kinase by touch, cold, and water stress in *Arabidopsis thaliana*. Proc Natl Acad Sci USA 93:765–769

Mizoguchi T, Ichimura K, Shinozaki K (1997) Environmental stress response in plants: the role of mitogen-activated protein kinases. Trends Biotechnol 15:15–19

Pöpping B, Gibbons T, Watson MD (1996) The *Pisum sativum* MAP kinase homologue (PsMAPK) rescues the *Saccharomyces cerevisiae hog1* deletion mutant under conditions of high osmotic stress. Plant Mol Biol 31:355–363

Romeis T, Piedras P, Zhang S, Klessig DF, Hirt H, Jones J (1999) Rapid Avr9- and Cf-9-dependent activation of MAP kinases in tobacco cell cultures and leaves: convergence in resistance gene, elicitor, wound and salicylate responses. Plant Cell 11:273–297

Ryals JA, Neuenschwander UH, Willits MG, Molina A, Steiner H-Y, Hunt MD (1996) Systemic acquired resistance. Plant Cell 8:1809–1819

Seger R, Krebs EG (1995) The MAPK signaling cascade. FASEB J 9:726–735

Seo S, Okamoto M, Seto H, Ishizuka K, Sano H, Ohashi Y (1995) Tobacco MAP kinase: a possible mediator in wound signal transduction pathways. Science 270:1988–1992

Stafstrom JP, Altschuler M, Anderson DH (1993) Molecular cloning and expression of a MAP kinase homologue from pea. Plant Mol Biol 22:83–90

Stratmann JW, Ryan CA (1997) Myelin basic protein kinase activity in tomato leaves is induced systemically by wounding and increases in response to systemin and oligosaccharide elicitors. Proc Natl Acad Sci USA 94:11085–11089

Subramaniam R, Després C, Brisson N (1997) A functional homolog of mammalian protein kinase C participates in the elicitor-induced defense response in potato. Plant Cell 9:653–664

Suzuki K, Shinshi H (1995) Transient activation and tyrosine phosphorylation of a protein kinase in tobacco cells treated with a fungal elicitor. Plant Cell 7:639–647

Suzuki K, Fukuda Y, Shinshi H (1995) Studies on elicitor-signal transduction leading to differential expression of defense genes in cultured tobacco cells. Plant Cell Physiol 36:281–289

Usami S, Banno H, Ito Y, Nishihama R, Machida Y (1995) Cutting activates a 46-kilodalton protein kinase in plants. Proc Natl Acad Sci USA 92:8660–8664

Viard M-P, Martin F, Pugin A, Ricci P, Blein J-P (1994) Protein phosphorylation is induced in to-
    bacco cells by the elicitor cryptogein. Plant Physiol 104:1245–1249
Wasternack C, Parthier B (1997) Jasmonate-signalled plant gene expression. Trends Plant Sci 2:
    302–307
Wei Z-M, Laby RJ, Zumoff CH, Bauer DW, He SY, Collmer A, Beer SV (1992) Harpin, elicitor of
    the hypersensitive response produced by the plant pathogen Erwinia amylovora. Science
    257:85–88
Wilson C, Eller N, Gartner A, Vicente O, Heberle-Bors E (1993) Isolation and characterization of
    a tobacco cDNA clone encoding a putative MAP kinase. Plant Mol Biol 23:543–551
Wilson C, Anglmayer R, Vicente O, Heberle-Bors E (1995) Molecular cloning, functional expres-
    sion in Escherichia coli, and characterization of multiple mitogen-activated protein kinases
    from tobacco. Eur J Biochem 233:249–257
Wilson C, Voronin V, Touraev A, Vicente O, Heberle-Bors E (1997) A developmentally regulated
    MAP kinase activated by hydration of tobacco pollen. Plant Cell 9:2093–2100
Yang Y, Shah J, Klessig DF (1997) Signal perception and transduction in plant defense responses.
    Gene Dev 11:1621–1639
Yu LM (1995) Elicitins from Phytophthora and basic resistance in tobacco. Proc Natl Acad Sci
    USA 92:4088–4094
Yu LM, Lamb CJ, Dixon RA (1993) Purification and biochemical characterization of proteins
    which bind to the H-box cis-element implicated in transcriptional activation of plant defense
    genes. Plant J 3:805–816
Zhang S, Klessig DF (1997) Salicylic acid activates a 48 kDa MAP kinase in tobacco. Plant Cell 9:
    809–824
Zhang S, Klessig DF (1998a) Resistance gene N-mediated de novo synthesis and activation of a to-
    bacco MAP kinase by tobacco mosaic virus infection. Proc Natl Acad Sci USA 95:7433–7438
Zhang S, Klessig DF (1998b) The tobacco wounding-activated MAP kinase is encoded by SIPK.
    Proc Natl Acad Sci USA 95:7225–7230
Zhang S, Du H, Klessig DF (1998) Activation of the tobacco SIP kinase by both a cell wall-derived
    carbohydrate elicitor and purified proteinaceous elicitins from Phytophthora spp. Plant Cell
    10:435–449

# Receptor-Mediated MAP Kinase Activation in Plant Defense

Heribert Hirt[1] and Dierk Scheel[2]

**Summary:** Plants mount a complex array of defense reactions in response to attack by pathogens. Initiation of these events depends on perception and signal transduction of elicitors, which are plant-derived or pathogen-derived signals, that give rise to transcriptional activation of defense-related genes as well as to changes in activities of enzymes involved in cell wall reinforcement and oxygen radical formation. An oligopeptide, identified within a 42 kDa glycoprotein elicitor from *Phythophthora sojae,* activates in parsley cells typical plant defense reactions, enabling researchers to study plant-pathogen interaction at the single cell level. The oligopeptide elicitor was found to be necessary and sufficient to stimulate a complex defense response in parsley cells, comprising $H^+/Ca^{2+}$ influxes, $K^+/Cl^-$ effluxes, activation of a mitogen-activated protein (MAP) kinase, an oxidative burst, defense-related gene activation, and phytoalexin formation.

# 1
# Introduction

Plants are resistant towards the majority of potentially pathogenic organisms in their environment, since they are not the appropriate host plant (non-host or species resistance). Only relatively few pathogens managed to establish true host/pathogen interactions with the plant being susceptible and the pathogen virulent. Within such systems of basic compatibility, host or cultivar resistance exists, in which host plant cultivars have developed specific resistances against distinct pathogen races (Heath 1997). The incompatibility of these interactions relies on specific avirulence genes of the pathogen and resistance genes of the host plant according to the gene-for-gene hypothesis (Flor 1955; De Wit 1992). The corresponding physiological model proposes that avirulence genes directly or indirectly encode cultivar-specific elicitors that bind to plant receptors encoded by the matching resistance gene (De Wit 1992). Although several pairs

[1] Institute of Microbiology und Genetics, Dr.-Bohr-Gasse 9, A-1030 Wien, Austria
[2] Institute of Plant Biochemistry, Weinberg 3, D-06120 Halle/ Saale, Germany

Results and Problems in Cell Differentiation, Vol. 27
Hirt (Ed.): MAP Kinases in Plant Signal Transduction
© Springer-Verlag Berlin Heidelberg 2000

of avirulence and resistance genes have been cloned, ligand/receptor interactions of the corresponding proteins remain to be demonstrated for most cases (Ellis and Jones 1998). Non-host resistant plants apparently also recognize potentially pathogenic organisms through receptors on their surface that specifically bind elicitors not encoded by avirulence genes, but derived from pathogen or plant surface structures (Ebel and Scheel 1997). Similar, if not identical signaling and defense elements appear to be employed by plant cells in host and non-host resistance (Ebel and Scheel 1997), suggesting that coupling of receptor-mediated recognition processes with inducible defense reactions is an early event in plant evolution (Heath 1997). Functional analysis of these signal transduction chains is required to understand interorganismic communication and may help to design novel plant protection strategies.

## 2
## *Phytophthora sojae* Elicitor-Induced Responses of Parsley Cells: A Model System for Plant-Pathogen Interaction

Cultured parsley cells and a glycoprotein elicitor from hyphal cell walls of the plant pathogenic oomycete, *Phytophthora sojae*, have been used as an experimental system to investigate the molecular mechanisms underlying non-host recognition and signaling (Parker et al. 1991). Within the glycoprotein, an oligopeptide fragment of 13 amino acids was found to be necessary and sufficient for elicitor activity (Sacks et al. 1995). This oligopeptide fragment, Pep-13, stimulates a multicomponent defense response in cultured parsley cells (Table 1) which is identical to the reaction to the whole glycoprotein and very similar to the response of intact plants to infection with the fungus (Fig. 1; Jahnen and Hahlbrock 1988; Parker et al. 1991; Nürnberger et al. 1994a). Since this elicitor, in addition, is located in hyphal cell walls in infected plant tissue (Hahlbrock et al. 1995), it is believed to act as a recognition signal in this non-host plant/pathogen interaction.

**Table 1.** Oligopeptide-induced responses of parsley cells

| Response | Time of initiation |
| --- | --- |
| Ion fluxes through the plasma membrane | 2–3 min |
| Oxidative burst | 4–5 min |
| Jasmonate accumulation | 4–5 min |
| *In vivo* protein phosphorylation | 1–10 min |
| Defense gene activation | 15–30 min |
| mRNA accumulation | 0.5–1 h |
| Protein accumulation | 1–2 h |
| Phytoalexin accumulation | 4–5 h |

**Fig. 1.**
Temporal and spatial pattern
of plant defense reactions
against pathogens

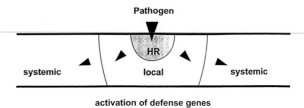

**Reactions of infected tissue**
browning                                         immediately
oxidative burst
callose deposition
cell death

**Local gene activation**
phenylpropanoid biosynthesis          early
defense proteins
phytoalexin biosynthesis

**Systemic gene activation**
glucanases                                       delayed
chitinases
defense proteins

# 3
# Pep-13 Elicitor Binds to a 100 kDa Plasma Membrane Receptor

Binding experiments with radio-iodinated Pep-13 revealed the existence of a
high-affinity binding site at plasma membranes of parsley cells (Nürnberger et
al. 1994b). All responses of the plant cells to Pep-13 appear to be mediated
through this binding site, since the binding affinities of different Pep-13 derivatives correlate with their elicitor activities. Cross-linking experiments and
partial purification of the binding site suggest that a 100-kDa plasma membrane protein is the functional receptor (Nürnberger et al. 1995; Nennstiel et al.
1998).

# 4
# Signal Transduction Events
# of Pep-13 Elicitor-Induced Defense Responses

Early elements of the plant response are ion fluxes across the plasma membrane, changes in protein phosphorylation patterns, the production of reactive

oxygen species (oxidative burst) and jasmonate synthesis (T. Kroj and D. Scheel, unpubl.) followed by defense gene activation and phytoalexin accumulation (Table 1; Scheel 1998). Activation of the elicitor-responsive ion channels, the most rapid reaction, has been found to be necessary and sufficient for all other reactions of the plant cells (Jabs et al. 1997). Patch clamp analysis of intact parsley protoplasts was employed to electrophysiologically characterize one Pep-13-responsive $Ca^{2+}$ channel in the plasma membrane which appears to be indirectly regulated through the Pep-13 receptor (Zimmermann et al. 1997). Inhibition of this ion channel with $La^{3+}$ or $Gd^{3+}$ blocks all cellular responses to the elicitor. Cytosolic $Ca^{2+}$ levels increase transiently in Pep-13-treated aequorin-expressing parsley cells with kinetics similar to elicitor-stimulated $Ca^{2+}$ influxes (B. Blume and D. Scheel, unpubl.).

Inhibition of elicitor-stimulated production of reactive oxygen species with diphenylene iodonium (DPI), an inhibitor of the mammalian NADPH oxidase, blocks jasmonate accumulation, defense gene activation and phytoalexin synthesis without affecting ion fluxes, cell viability and constitutive gene expression (Jabs et al. 1997; Ligterink et al. 1997). Mimicking the oxidative burst in the absence of elicitor by addition of appropriate amounts of $KO_2$ to the medium of cultured parsley cells stimulates phytoalexin accumulation but not ion fluxes (Jabs et al. 1997) and jasmonate production. $H_2O_2$ either added directly to or generated in the culture medium by glucose and glucose oxidase does not stimulate any defense response.

## 5
## Pep-13 Elicitor Induces Activation of a MAP Kinase

Elicitor signal transduction in parsley cells involves $Ca^{2+}$-dependent transient changes in protein phosphorylation, suggesting the participation of protein kinases in defense gene activation (Dietrich et al. 1990). When parsley cells were treated with Pep-13, a protein kinase that phosphorylated myelin basic protein (MBP) was activated within 5 min (Ligterink et al. 1997). From its relative mobility on SDS-polyacrylamide gels, the apparent molecular mass of this enzyme was estimated to be approximately 45 kDa, reminiscent of known plant MAP kinases. Using antisera that recognize and distinguish between different alfalfa MAP kinases (Jonak et al. 1996), a protein kinase was identified which was activated by Pep-13 elicitor (Ligterink et al. 1997).

Since the antiserum specifically recognized the elicitor-responsive MAP kinase from parsley, a radiolabeled fragment of the alfalfa MAP kinase gene was used to screen a parsley cDNA library. A clone was isolated which encodes a protein of 371 amino acids, corresponding to a molecular mass of 43 kDa and showing high similarity to the MAP kinases from *Arabidopsis*, ATMPK3 (83%; Mizoguchi et al. 1996), alfalfa, MMK4 (81%; Jonak et al. 1996), and tobacco, WIPK (83%; Seo et al. 1995). The overall structure of the parsley, tobacco,

*Arabidopsis* and alfalfa kinases is highly conserved, suggesting that these genes may perform similar functions in the different species. RNA gel blot analysis of cultured parsley cells showed increased levels of kinase transcript within 30 min after elicitor treatment. The parsley gene was expressed as a fusion protein with glutathione S-transferase (GST) in *Escherichia coli,* revealing autophosphorylation and kinase activity towards MBP. In immunoblots, the GST-MAP kinase fusion protein was recognized by the antiserum that monospecifically crossreacted with the elicitor-responsive protein kinase from cultured parsley cells. These results suggested that the cDNA isolated from parsley cells indeed encodes the elicitor-activated kinase detected in the activity and immunocomplex assays.

# 6
## MAP Kinase Activation Occurs Through the Pep-13 Receptor

In order to investigate whether Pep-13 activates this MAP kinase through the same receptor that is used for the induction of the other defense responses, MAP kinase activation was determined upon treatment of parsley cells with four different but structurally related elicitor oligopeptides. Two Pep-13 derivatives in which the second or the fifth amino acid (Pep-13A$_2$ and Pep-13A$_5$, respectively) had been replaced by alanine did not activate MAP kinase, whereas a derivative with an alanine substitution in position 12 (Pep-13A$_{12}$) was as active as Pep-13. These results correlate well with binding and elicitor studies (Nürnberger et al., 1994b) with the same Pep-13 derivatives, which showed that Pep-13A$_{12}$ competitively inhibits binding of Pep-13 to its receptor and elicits a normal pattern of defense reactions. In contrast, Pep-13A$_2$ and Pep-13A$_5$ were inactive in both assays, indicating that the Pep-13 receptor which is initiating the multicomponent defense response is also engaged in MAP kinase activation.

# 7
## MAP Kinase Activation Is Dependent on Ion Fluxes

Binding of Pep-13 to its receptor induces phytoalexin synthesis, defense gene activation, *in vivo* phosphorylation of proteins, and the producton of an oxidative burst, which all depend on the integrity of specific ion channels that are involved in mediating rapid ion fluxes across the cell membrane in response to elicitor (Dietrich et al. 1990; Nürnberger et al. 1994; Jabs et al. 1997; Zimmermann et al. 1997). To investigate whether MAP kinase activation also depended on the activity of these ion channels, parsley cells were incubated with the ion channel blocker, anthracene-9-carboxylate (A9C), which inhibits the elicitor-stimulated ion fluxes, blocking all subsequent defense responses (Jabs et al.

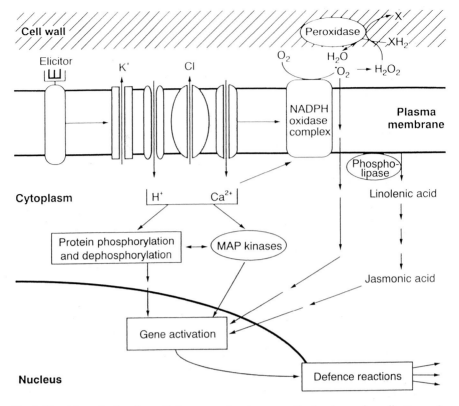

**Fig. 2.** Signal transduction from elicitor perception to gene activation in parsley cells. Recognition of the elicitor by its receptor, a 100-kDa plasma membrane protein, stimulates transient influx of $H^+$ and $Ca^{2+}$ and efflux of $K^+$ and $Cl^-$. These ion fluxes are necessary for changes in protein phosphorylation mediated by various kinases and phosphatases including MAP kinase, which is translocated into the nucleus. Ion fluxes are also a prerequisite for activation of an NADPH oxidase resulting in the production of reactive oxygen species (oxidative burst) which are responsible for peroxidase-mediated crosslinking of cell wall components and contribute to the induction of defense-related genes and synthesis of phytoalexins

1997). Under these conditions, Pep13 activation of the MAP kinase was completely inhibited, indicating that ion channel activation was also necessary for this reaction (Ligterink et al. 1997). Amphotericin B (Amph), which mimics elicitor-induced ion fluxes and thereby induces the full set of defense responses (Jabs et al. 1997), also activates the MAP kinase in the absence of elicitor (Ligterink et al. 1997). Activation of MAP kinase, ion fluxes and the oxidative burst (Jabs et al. 1997) by amphotericin B all occur after a delay of about 30 min. Thus, MAP kinase activation depends on the state of specific ion channels, and activation of these channels is necessary and sufficient for MAP kinase activation as it is for the induction of the other elicitor responses in this system.

# 8
## MAP Kinase Activation Is Independent of Oxidative Burst

The elicitor-stimulated production of reactive oxygen species is thought to be catalyzed by an NADH- or NADPH-oxidase that is inhibited by DPI (Mehdy et al. 1996). In elicitor-treated parsley cells, DPI blocked the oxidative burst, defense gene activation, and phytoalexin accumulation without affecting ion fluxes (Jabs et al. 1997). Together with the results from gain-of-function experiments with $KO_2$ which stimulated phytoalexin production in the absence of elicitor, this placed the oxidative burst downstream of the ion channels within the elicitor signal transduction cascade (Jabs et al. 1997). Pep-13 activation of the MAP kinase was not inhibited by DPI, indicating that this kinase acts either upstream or independently of the oxidative burst (Ligterink et al. 1997).

# 9
## Activation of MAP Kinase Is Correlated with Its Nuclear Translocation

Certain MAP kinases are translocated into the nucleus upon activation, where they may catalyze phosphorylation of transcription factors and thereby regulate gene transcription (Chen et al. 1992; Sanghera et al. 1992; Lenormand et al. 1993). The subcellular location of the elicitor-responsive MAP kinase was determined with antiserum in immunofluorescence microscopy before and after treating parsley cells with elicitor. Within 3–10 min after elicitor treatment, MAP kinase is translocated into the nucleus (Ligterink et al. 1997). Since no nuclear localization signal is present in MAP kinases, translocation of the activated kinase into the nuclear compartment may be initiated by its interaction with another protein, perhaps a transcription factor. In parsley, several elicitor-responsive genes have been identified and have led to the identification of *cis*-acting elements and transcription factors that may be involved in mediating pathogen-induced transcription (Rushton et al. 1996). Although it has not yet been shown that phosphorylation of these transcription factors is responsible for elicitor-induced transcription of pathogenesis-related genes, the elicitor-activated MAP kinase that relocates into the nucleus upon activation might link cytosolic signal transduction to nuclear activation of plant defense genes.

Taken together, our results demonstrate a causal relationship between early and late elicitor reactions, establish a sequence of signaling events from receptor-mediated activation of ion channels through MAP kinase activation, the oxidative burst, and defense gene activation to phytoalexin production, and suggest that within this signal transduction chain the superoxide anion radical rather than $H_2O_2$ triggers defense gene activation and phytoalexin accumulation.

*Acknowledgements.* Our work was supported by the Austrian National Bank (project 6159), Austrian Science Foundation (project P11729-GEN and P12188-GEN), Deutsche Forschungsgemeinschaft (Sche 235/3), the European Community (ERBBIO4CT960101) and the Fonds der chemischen Industrie.

# References

Chen R-H, Sarnecki C, Blenis J, et al. (1992) Nuclear localization and regulation of ERK- and RSK-encoded protein kinases. Mol Cell Biol 12:915–927

De Wit PJGM (1992) Molecular characterization of gene-for-gene systems in plant-fungus interactions and the application of avirulence genes in control of plant pathogens. Annual Reviews of Phytopathology 30:391–418

Dietrich A, Mayer JE, Hahlbrock K (1990) Fungal elicitor triggers rapid, transient, and specific protein phosphorylation in parsley cell suspension cultures. J Biol Chem 265:6360–6368

Ebel J, Scheel D (1997) Signals in host-parasite interactions. In: Carroll GC, Tudzynski P (eds) Plant Relationships, Part A, Springer, Berlin Heidelberg New York, V:85–105

Ellis J, Jones D (1998) Structure and function of proteins controlling strain-specific pathogen resistance in plants. Curr Opin Plant Biol 1:288–293

Flor HH (1955) Host-parasite interactions in flax rust – Its genetics and other implications Phytopathology 45:680–685

Hahlbrock K, Scheel D, Logemann E, et al. (1995) Oligopeptide elicitor-mediated defense gene activation in cultured parsley cells. Proc Nat Acad Sci USA 92:4150–4157

Heath MC (1997) Evolution of plant resistance and susceptibility to fungal parasites. In: Carroll GC, Tudzynski P (eds) Plant Relationships, Part A, Springer, Berlin Heidelberg New York, V: 257–276

Jabs T, Tschöpe M, Colling C, et al. (1997) Elicitor-stimulated ion fluxes and $O_2^-$ from the oxidative burst are essential components in triggering defense gene activation and phytoalexin synthesis in parsley. Proc Natl Acad Sci USA 94:4800–4805

Jahnen W, Hahlbrock K (1988) Cellular localization of nonhost resistance resistance reactions of parsley (*Petroselinum crispum*) to fungal infection. Planta 173:197–204

Jonak C, Kiegerl S, Ligterink W, et al. (1996) Stress signaling in plants: A mitogen-activated protein kinase pathway is activated by cold and drought. Proc Natl Acad Sci USA 93: 11274–11279

Lenormand P, Sardet C, Pages G, et al. (1993) Growth factors induce nuclear translocation of MAP kinases ($p42^{mapk}$ and $p44^{mapk}$) but not of their activator MAP kinase kinase ($p45^{mapkk}$) in fibroblasts. J Cell Biol 122:1079–1088

Ligterink W, Kroj T, zur Nieden U, et al. (1997) Receptor-mediated activation of a MAP kinase in pathogen defense of plants. Science 276 (5321):2054–2057

Mehdy MC, Sharma YK, Sathasivan K, et al. (1996) The role of activated oxygen species in plant disease resistance. Phys Plant 98:365–374

Mizoguchi T, Irie K, Hirayama T, et al. (1996) A gene encoding a mitogen-activated protein kinase kinase kinase is induced simultaneously with genes for a mitogen-activated protein kinase and an S6 ribosomal protein kinase by touch, cold, and water stress in *Arabidopsis thaliana*. Proc Natl Acad Sci USA 93:765–769

Nennstiel D, Scheel D, Nürnberger T (1998) Characterization and partial purification of an oligopeptide elicitor receptor from parsley (*Petroselinum crispum*). FEBS Lett 431:405–410

Nürnberger T, Colling C, Hahlbrock K, et al. (1994a) Perception and transduction of an elicitor signal in cultured parsley cells. Biochem Soc Symp 60:173–182

Nürnberger T, Nennstiel D, Hahlbrock K, et al. (1994b) High affinity binding of a fungal oligopeptide elicitor to parsley plasma membranes triggers multiple defense responses. Cell 78: 449–460

Nürnberger T, Nennstiel D, Jabs T, et al. (1995) Covalent cross-linking of the *Phytophthora megasperma* oligopeptide elicitor to its receptor in parsley membranes. Proc Natl Acad Sci USA 92:2338–2342

Parker JE, Schulte W, Hahlbrock K, et al. (1991) An extracellular glycoprotein from *Phytophthora megasperma* f. sp. *glycinea* elicits phytoalexin synthesis in cultured parsley cells and protoplasts. Mol Plant-Microbe Interact 4:19–27

Rushton PJ, Torres JT, Parniske M, et al. (1996) Interaction of elicitor-induced DNA-binding proteins with elicitor response elements in the promoters of parsley PR1 genes. EMBO J 15:5690–5700

Sacks WR, Nürnberger T, Hahlbrock, et al. (1995) Molecular characterization of nucleotide sequences encoding the extracellular glycoprotein elicitor from *Phytophthora megasperma*. Mol Gen Genet 246:45–55

Sanghera JS, Peter M, Nigg EA, et al. (1992) Immunological characterization of avian MAP kinases: Evidence for nuclear localization. Mol Cell Biol 3:775–787

Scheel D (1998) Resistance response physiology and signal transduction. Curr Opin Plant Biol 1:305–310

Seo S, Okamoto M, Seto H, et al. (1995) Tobacco MAP kinase: a possible mediator in wound signal transduction pathways. Science 270:1988–1992

Zimmermann S, Nürnberger T, Frachisse J-M, et al. (1997) Receptor-mediated activation of a plant $Ca^{2+}$-permeable ion channel involved in pathogen defense. Proc Natl Acad Sci USA 94:2751–2755

# Regulation of Cell Division and the Cytoskeleton by Mitogen-Activated Protein Kinases in Higher Plants

László Bögre[1], Ornella Calderini[1,3] Irute Merskiene[1], and Pavla Binarova[2]

**Summary:** The microtubule-associated protein 2 kinase (MAP2-kinase), now better known as mitogen-activated protein kinase (MAPK), was initially discovered in association with the cytoskeleton, and was later also implicated in cell division. The importance of mitogenic stimulation in plant development roused interest in finding the plant homologues of MAPKs. However, data on plant MAPKs in cell division are rather sparse and fragmentary. Therefore we place the available information on cell cycle control of MAPKs in plants into a broader context. We discuss four aspects of cell division control: cell proliferation and the $G_1/S$-phase transition, $G_2$-phase and mitosis, cytokinesis, and cytoskeletal reorganisation. Future work will reveal to what extent plants use signalling pathways that are similar or different to those of animal or yeast cells in regulating cell divisions.

## 1
## Regulation of the $G_1/S$ Transition: The Animal Paradigm

Molecular events involved in $G_1$ control in animal cells are well established (Fig. 1). First, we summarise the present knowledge on $G_1$-regulation in animal cells (Fig. 1), referring readers to recent reviews for further details and references (Treisman 1996; Marshall 1996; Bottazzi and Assoian 1997; Robinson and Cobb 1997; Thomas and Hall 1997; Peterson and Schreiber 1998; Rommel and Hafen 1998).

Various types of growth factors (GF) activate tyrosine kinase receptors (RTK) by inducing their oligomerisation and phosphorylation. Receptor activation leads to the conversion of a small GTPase, Ras, from its inactive (Ras

[1] Vienna Biocenter, Institute of Microbiology and Genetics, University of Vienna, Dr. Bohrgasse 9, 1030 Vienna, Austria
[2] Institute of Microbiology, Academy of Sciences of the Czech Republic, Videnska 1083, 142 20 Prague, Czech Republic
[3] Instituto di Ricerche sul Miglioramento Genetico Piante Foraggere CNR, via Madonna Alta 130, 06128 Perugia, Italy

Results and Problems in Cell Differentiation, Vol. 27
Hirt (Ed.): MAP Kinases in Plant Signal Transduction
© Springer-Verlag Berlin Heidelberg 2000

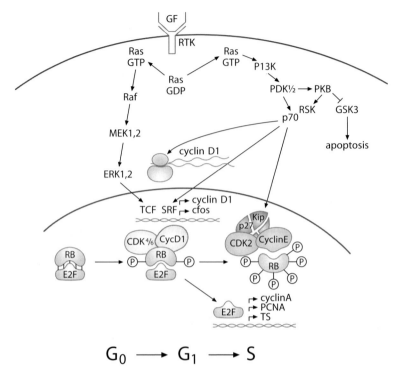

**Fig. 1.** $G_1$ control in animal cells. Growth factors *GF* bind to and activate receptor tyrosine kinases *RTK* which stimulate the exchange of GDP to GTP bound to Ras. Ras activates two independent pathways. It activates Raf, which is a MAP kinase kinase kinase, which in turn activates two related MAP kinase kinases, *MEK1* and *MEK2* and further on, two related MAP kinases, *ERK1* and *ERK2*. Upon activation, *ERK1* and *ERK2* translocate to the nucleus and phosphorylate the ternary complex factor *TCF*, required for the transcriptional activation of early response genes, such as c-fos or cyclin D1. In another pathway the GTP-bound Ras activates the phosphoinositide-3-kinase *PI3K*. The released phosphoinositides activate two related protein kinases, the phosphoinositide-dependent protein kinase 1 and 2 *PDK1* and *PDK2*. *PDK1* and *PDK2* are the first protein kinases in this cascade which activate further protein kinases, protein kinase B *PKB* and p70 ribosomal kinase $p70^{Rsk}$. PKB inhibits the glycogen synthase kinase 3 *GSK3* and thereby apoptosis. The $p70^{Rsk}$ performs multiple functions: it phosphorylates the partner transcription factor of *TCF*, the serum response factor *SRF*, enhancing the translation of certain mRNAs, such as cyclin D1, and induces the proteolysis of a cyclin-dependent kinase inhibitor, $p27^{Kip1}$. All these events culminate in a master switch operating during the $G_0$ to $G_1$ to S-phase transitions, releasing the retinoblastoma protein *Rb*-bound *E2F*. *E2F* is a transcription factor for S phase-specific genes, such as cyclin A, proliferation cell nuclear antigen *PCNA* or thymidilate synthase *TS*

GDP) to its active (Ras GTP) form. The pathway bifurcates at Ras into two signalling cascades that co-operate in mitogenic stimulation. In one cascade, Ras binds and activates Raf which leads to the activation of the MAPK pathway through the phosphorylation of MEK1 or MEK2 and subsequently ERK1 or

ERK2. The MAPKs ERK1 and ERK2 translocate to the nucleus upon mitogenic stimulation and phosphorylate transcription factors eg. TCF, which leads to the expression of immediate early genes, such as c-fos, as well as the transcription of the $G_1$ cyclin, cyclin D1.

The other pathway involves the activation of PI3K by Ras and thereby the production of phospholipids (PIPs). PIPs activate two related protein kinases, PDK1 and PDK2, which phosphorylate and thereby activate PKB, another membrane-bound protein kinase. PKB has a dual role. It inactivates GSK3 kinase by phosphorylation, thus inhibiting apoptosis, an important step toward immortalised proliferation, and together with PDK1 and PDK2 it phosphorylates, and in this case activates, $p70^{Rsk}$, which also performs multiple functions. $p70^{Rsk1}$ phosphorylates the transcription factor SRF which, together with TCF, is required for the expression of immediate-early genes such as the proto-oncogene c-fos and for the expression of the $G_1$ cyclin, cyclin D1. Another target of $p70^{Rsk}$ is the ribosomal protein S6, the phosphorylation of which enhances the translation of mRNAs containing a 5' tract of polypyrimidine, such as the cyclin D1 mRNA. $p70^{Rsk}$ is also known to phosphorylate the cyclin-dependent kinase (CDK) inhibitor $p27^{Kip1}$ and in so doing targets it for degradation.

The increased transcription and translation of cyclin D1 and the proteolysis of $p27^{Kip1}$ leads to the subsequent activation of CDK4- or CDK6-cyclin D1 and CDK2-cyclin E complexes. All these events culminate in the phosphorylation of a master switch of $G_1$ control, the retinoblastoma protein (Rb), first by the CDK4- or CDK6-cyclin D1 complexes and then by the CDK2-cyclin E complex. Hyperphosphorylated Rb does not bind any more to the transcription factor E2F, which induces the expression of genes required for DNA synthesis, such as thymidilate synthase (TS), proliferating cell nuclear antigen (PCNA), and cyclin A.

# 2
# The Plant Players of $G_1$ Control

Plant hormones have a global effect in promoting (cytokinin, gibberellic acid, auxin) or inhibiting (ethylene, abscisic acid) cell division. Local developmental regulators also exist in plants, which influence cell division activities in the meristem by cell to cell communication (Meyerowitz 1997).

The isolation and molecular characterisation of mutants and the cloning of genes homologous to signalling components in other systems is beginning to unravel some of the signalling mechanisms in plants. Here we list plant receptors and signalling components, and sketch some of the possible connections to the cell cycle machinery.

Cytokinins were discovered as substances promoting cell division in combination with auxin. By the isolation of mutants insensitive to cytokinin, the signal transduction pathway is beginning to be understood (Kakimoto 1998). Unlike the ethylene pathway, where a histidine kinase receptor is connected to a MAPK module (Solano and Ecker 1998), cytokinin signalling appears to be similar to the canonical bacterial two component system, where the histidine-kinase receptor directly or indirectly activates a response regulator, which is a transcription factor. Using a reverse-genetic approach, a seven-transmembrane-receptor has also been implicated in cytokinin signalling (Plakidou-Dymock et al. 1998). Cytokinins were shown to induce the expression of the $G_1$ cyclin, cycD3, in *Arabidopsis* (Fuerst et al. 1996).

Another growth promoting hormone is gibberellic acid. The receptor for it is not known at present, but this signalling pathway is thought to contain heterotrimeric G proteins, which potentially could be linked to a MAPK pathway (Jones et al. 1998). In one report, however, the downregulation of a MAPK transcript was reported when aleurone cells were treated with gibberellic acid (Huttly and Phillips 1995). In the same study the transcription of another gene, *aspk10*, with significant homology to p70$^{Rsk}$ and p90$^{Rsk}$, was induced upon gibberellic acid treatment. *Atatpk1*, also belonging to the p70$^{Rsk}$ group of protein kinases, was found to be abundant in dividing tissues, which is consistent with a role in cell division control (Zhang et al. 1994).

MAPKs have also been implicated in the signalling of the growth promoting hormone, auxin. A plant MEKK, NPK1, was found to inhibit auxin-induced gene expression (Kovtun et al. 1998), while treatment of cells with auxin transiently activated a plant MEK and MAPK (Mizoguchi et al. 1994). However, recently it was shown that the activation of MAPK in this experiment could be attributed to weak acids rather than to the physiological concentration of auxin (Tena and Renaudin 1998). Genes identified by the analysis of auxin-signalling mutants, such as aux1, show striking similarities to components of the protein degradation machinery, some of which are involved in $G_1$ control in yeast (Walker and Estelle 1998). The involvement of aux1 in the degradation of a CDK inhibitor could provide a mechanism by which auxin stimulates cell division.

Abscisic acid, a growth-inhibitory plant hormone, also activates a MAPK pathway in barley aleurone cells (Knetsch et al. 1996). The abscisic acid signalling mutants abi1 and abi2 code for a PP2C type phosphatase. This class of phosphatases was shown to inactivate stress-signalling MAPK pathways in yeasts, mammals, and plants, but it is not known whether ABI1 and ABI2 play similar roles (Meskiene et al. 1998). Abscisic acid signalling impinges on the transcription of specific genes. The induction of the expression of a CDK inhibitor, the *Arabidopsis* ICK1, is compatible with the growth inhibitory function of abscisic acid (Wang et al. 1998).

Members of the plant receptor-like protein kinase family (RLK) are structurally similar to growth factor receptors of animal cells (Becraft 1998). RLKs are membrane-bound proteins with extracellular ligand-binding and intracellular serine/threonine protein kinase domains. A RLK was identified through analysis of mutations in the clav1 locus, which appears to function in restricting cell division to the meristem, or by promoting the differentiation of cells which are leaving the proliferation zone in the meristem. This class of receptors is involved in short-range cell to cell signalling and could provide a link between developmental cues and cell division activities within the meristem. However, the ligands and the downstream signalling events of RLKs are largely unknown.

Promotion or inhibition of growth might be exerted by the differential effect of plant hormones on positive and negative regulators of cell division. The growth-inhibitory hormone abscisic acid induces the expression of a CDK inhibitor, ICK1, while the growth-promoting auxin might target a CDK inhibitor for degradation. Cytokinin might induce cell division by influencing the expression of the cycD3 $G_1$ cyclin while gibberellic acid might promote the translation of $G_1$ cyclins. Changes in CDK inhibitor and $G_1$ cyclin levels could operate the molecular switch of $G_1$ control composed of Rb and E2F, as has been found in animal cells (Murray 1997).

Two studies aimed at investigating whether MAPKs are involved in $G_1$ phase control have been published. In one experiment cultured tobacco BY2 cells were arrested in the $G_1$ phase of the cell cycle by omitting phosphate from the medium (Wilson et al. 1998). After re-feeding with phosphate, the cells resumed proliferation. Entry into the cell cycle correlated with a rapid and transient activation of a tobacco MAPK related to the alfalfa MMK1, several hours before cells entered S phase (6–8 hours later). Nucleolin, a protein involved in rRNA biogenesis serves as a molecular marker for monitoring the entry of cells into a metabolically active state of proliferation. In alfalfa cells nucleolin was found to be expressed 4 h after the re-addition of phosphate, which is shortly before S-phase (Bögre et al. 1996). $G_1$ cyclins are key regulators of entry into the cell cycle in animal, yeast, as well as in plant cells. When phosphate-starved $G_1$-arrested cells resume cell division, $G_1$ cyclins are expressed 6 h after re-feeding with phosphate at the $G_1/S$ transition (Dahl et al. 1995). In animal cells the activation of MAPK is directly linked to the induction of $G_1$ cyclin expression (Lavoie et al. 1996). The efficient activation of cell proliferation requires sustained activation of the MAPK pathway (Zhu and Assoian 1995). The very transient activation profile and the considerable time gap between the activation of MAPK and the induction of nucleolin and $G_1$ cyclin expression makes it uncertain whether the MAPK activity after re-feeding phosphate is directly related to the entry into the cell cycle. In another report, however, the activation of Ras in quiescent animal cells treated with serum was found to be biphasic, the first activation of Ras was achieved rapidly after serum addition, while the second activation occurred some 5 h later in the mid-$G_1$ phase (Taylor and Shalloway 1996). Interestingly, only the early phase of Ras activation correlates with activation of the Raf/MAPK pathway, while at later times high levels of Ras-GTP occurred without significant MAPK activity.

In another experiment, tobacco cells were arrested in the $G_1$ phase by omitting auxin from the medium. Upon auxin addition cells resumed semi-synchronous cell division with similar kinetics to phosphate starvation, entering S phase after 8 h and mitosis at around 15–18 h. Similarly, a rapid and transient activation of kinase activity was found, which phosphorylated myelin basic protein, peaking sharply 10 min after the re-addition of auxin to the medium (Mizoguchi et al. 1994). A MEK was also activated with the same kinetics in these experiments. The rapid and transient nature of this activity, in contrast to the resumption of cell division, throws doubt on a direct relationship between this MAPK activation and auxin-induced cell division.

Many components in signalling have been identified, but linking signal transduction to cell proliferation remains a challenge. The present biochemical approaches must be complemented with genetic tools in order to progress in this area.

## 3
## Regulation of $G_2$/M Transition: The Animal Paradigm

There is ample evidence for the involvement of MAPKs in $G_1$ control, but their role in the $G_2$/M transition in somatic animal cells is still not clear. The activation of ERK1 and ERK2 MAPKs not only in the $G_1$ but also during the $G_2$ phase indicates a function in mitosis (Tamemoto et al. 1992; Heider et al. 1994; Philipova and Whitaker 1998; Chiri et al. 1998). Some of the upstream regulators, such as the proto-oncogenes Raf (Laird et al. 1995; Patham et al. 1996) and Src (Roche et al. 1995), but not Ras (Taylor and Shalloway 1996), and some of the downstream targets of MAPKs, such as the c-fos transcription (Liu et al. 1994), are all activated during the $G_2$ phase in animal cells (Fig. 2). Using antibodies specific for activated MAPKs, ERK1 and ERK2 have been found to invade the nucleus in prophase and were detected on various mitotic structures, i.e. at kinetochores when chromosomes were in the process of alignment, in centrosomes during the entire period of mitosis, and in the midbody during cytokinesis (Shapiro et al. 1998; Zecevic et al. 1998). ERK1 and ERK2 were shown to phosphorylate a kinetochore-bound protein on a mitosis-specific phosphoepitope recognised by the 3F3/2 monoclonal antibody (Shapiro et al. 1998). This phosphorylation event was implicated in the checkpoint control for chromosome alignment (Campbell and Gorbsky 1995). The kinesin microtubule motor protein, CENP-E, is one of the best characterised mitotic substrates for ERK1 and ERK2. CENP-E is located on the kinetochore during metaphase and is involved in chromosome alignment (Liao et al. 1994; Wood et al. 1997, Zecevic et al. 1998). Together, these data suggest that ERK1 and ERK2 are involved in multiple steps during mitosis. However, the precise role for ERKs during mitosis is uncertain. The inhibition of their function by microinjection of ERK-specific antibodies or negative regulators of the MAPK pathway, such as the CL100 phosphatase, into mitotic somatic cells does not have a readily observable effect on normal cell cycle progression (Wang et al. 1997). In contrast, inactiva-

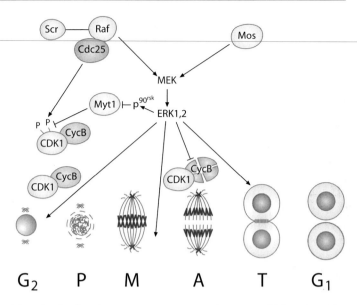

**Fig. 2.** Mitotic control in animal cells. There are two MEKKs, Raf or Mos, which activate the ERK1, ERK2 MAPK pathway in mitotic somatic cells or in oocytes, respectively. ERK1 and ERK2 have multiple roles during mitotic progression: $G_2$ phase $G_2$, prophase $P$, metaphase $M$, anaphase $A$, telophase $T$ and $G_1$ phase $G_1$. ERK1 and ERK2 activate the ribosomal kinase, p90[Rsk], which phosphorylates and thereby inactivates the Myt1 kinase. Myt1 kinase phosphorylates Cdk1 on the inhibitory Thr14 and Tyr15 sites. An activator of CDK1, the Cdc25 phosphatase, is recruited to Raf. Another possible function of ERK1 and ERK2 is to inhibit Cyclin B degradation

tion of ERKs seems to affect the checkpoint control for microtubule integrity and the alignment of chromosomes in metaphase (Minshull et al. 1994; Wang et al. 1997; Takenaka et al. 1997, 1998).

The involment of MAPKs during meiotic cell divisions in oocytes is more clear. ERKs are utilised to start oocyte maturation and meiosis I from $G_2$-arrested oocytes and to arrest cell cycle progression in a metaphase-like state at the end of meiosis II (Sagata 1997).

MAPKs activate mitotic CDKs through the action of p90[Rsk] which phosphorylates and thereby inactivates the Myt1 kinase (Palmer et al. 1998). The Myt1 kinase is a partner enzyme of the Wee1 kinase for phosphorylating CDK1 on the Thr14 and Tyr15 inhibitory sites. Inactivation of the CDK1 inhibitory kinase, Myt1, thereby activates CDK1. The Cdc25 phosphatase, which opposes the effect of Wee1 and Myt1, is recruited to and possibly activated by Raf (Galaktionov et al. 1995), providing other means to activate CDK1 (Figure 2). As CDK1, ERK1 and ERK2 are all activated simultaneously during mitosis and have overlapping substrate specificities (Kennelly and Krebs 1991), it is hard to separate the functions of these kinases.

# 4
# Plant Mitosis; Mechanically Different but the Regulation Might be the Same

In contrast to animal somatic cells, plant cells are capable of exiting the cell cycle and differentiating from both the $G_1$ and $G_2$ phases. Consequently, plant hormones and nutrients stimulate cell cycle progression at both control points (Van't Hof 1974). Though the downstream cell cycle targets are very different in the $G_1$ and $G_2$ phases, the same MAPK signalling pathway could be used, as has been shown in animal cells, where MAPKs are required for entry into the cell cycle at the $G_1$ phase in somatic cells or at the $G_2$ phase in oocytes.

That protein phosphorylation regulates plant mitosis is indicated by the prolonged metaphase in an experiment in which cells were treated with the broad-specificity protein kinase inhibitor K252-a (Wolniak and Larsen 1995).

At present there are only two examples of a possible involvement of plant MAPKs in $G_2$ control. The mRNA levels of an alfalfa MAPK, MMK1, were found to be more abundant in $G_2$-phase cells when the cell cycle was synchronised by a release from an aphidicolin block at the $G_1$/S boundary (Jonak et al. 1993). The changing MMK1 abundance might relate to some events of cytoskeleton reorganisation during cell cycle progression in $G_2$ cells, such as the formation of the preprophase band, which is a plant-specific microtubule structure formed in $G_2$ cells and which functions to position the plane of cell division. MMK1-related MAPKs in tobacco plants, however, were found to respond to environmental stimuli, such as the rehydration of pollen (Wilson et al. 1997) or wounding (Zhang and Klessig 1998). Perhaps, similarly to ERKs, MMK1 might perform multiple roles.

Two structurally similar plant MAPKs, the *Medicago* MAP kinase 3 (MMK3) and the *Nicotiana tabacum* FUS3-related kinase (NTF6), might regulate some aspects of cytokinesis (Calderini et al. 1998; Bögre et al. 1999). The MMK3 protein and its associated activity could be detected only in dividing cells. Though the MMK3 and NTF6 proteins are present throughout the cell cycle, both are transiently activated during mitosis. The entry into mitosis is required for the activation of these MAPKs. This was revealed by arresting cells in prophase with a CDK-specific drug, roscovitine. The timing of MMK3 and NTF6 activation was further dissected by synchronisation with the drug propyzamide, which depolymerises microtubules and blocks cells in metaphase. Both the alfalfa and tobacco kinases, similar to the animal ERKs (Tamemoto et al. 1992; Heider et al. 1994), were inactive in metaphase cells with depolymerised microtubules and became active after removing the drug as cells passed through anaphase and telophase. It is not simply the integrity of microtubules that influences the activity of MMK3 and NTF6. This was indicated by the reduced NTF6 activity when cells were treated with taxol. As opposed to propyzamide, taxol stabilises microtubules. The abnormally stable microtubules block cells in mitosis. This experiment suggested that it was not only the presence of intact microtubules but the passage through anaphase and telophase that was required for NTF6 activation. Another attempt to clarify the situation was to synchronise alfalfa cells by releasing them from a metaphase arrest and then depolymerising the microtubules with half-hour pulse treatments of amiprophosmethyl (APM) during anaphase and telophase. Inactivation of MMK3 was also apparent in this experiment. The transient nature of MAPK activation was further indicated by an experiment in which the time course of MMK3 and NTF6 activity was followed after APM application. This shows that both NTF6 and MMK3 are inactivated after 10 min. Both experiments still leave us with two interpretations: either intact microtubules are required during ana- and telophases, or the activation of MMK3 and NTF6 is transient and falls within the 10–30 minutes interval of the APM treatments.

In *Xenopus*, a MAPK was found to control the spindle assembly checkpoint and arrest cells in metaphase until chromosomes are aligned at the cell equator (Minshull et al. 1994; Takenaka et al. 1997, 1998). MMK3 and NTF6, however, are not active in metaphase arrested cells, when the spindle is depolymerised by a microtubule drug, and only become activated when the metaphase arrest is released and cells enter late anaphase. These data make it unlikely that MMK3 and NTF6 are involved in a checkpoint for controlling microtubule integrity.

MMK3 and NTF6 localisation in prophase is reminiscent of that of ERK1 and ERK2. Both ERKs, as well as MMK3, are found in the cytoplasm in interphase cells and invade the nucleus at the end of $G_2$ (Shapiro et al. 1998; Zecevic et al. 1998; Bögre et al. 1999). Though MMK3 and NTF6 are most active during ana- and telophase, some activation could already be found at prophase. MMK3, NTF6, as well as ERKs, were found at comparable locations in late mitosis. While MMK3 and NTF6 were found associated with the plant-specific phragmoplast in late anaphase and at the midplane of cell division in telophase cells, active ERK1 and ERK2 were detected at the midzone region in late anaphase and at the midbody during telophase and cytokinesis in animal cells (Shapiro et al. 1998; Zecevic et al. 1998). The fact that both ERKs as well as MMK3 behave similarly throughout mitosis with respect to activity and location suggests that these MAPKs could play similar roles in animal and plant mitosis, respectively.

The tobacco MEKK, NPK1, has also been implicated in mitosis (Banno et al. 1993). NPK1 protein levels increase as cells progress to mitosis (Machida et al. 1998). In this respect the regulation of NPK1 is more similar to MMK1 than MMK3, since MMK1 mRNA was also found to be more abundant in $G_2$ cells, while MMK3 and NTF6 mRNA and protein levels do not change during the cell cycle. A search for activators of NPK1 in a yeast screen led to the identification of kinesin-like proteins (Machida et al. 1998). ERK1 and ERK2 are also complexed with a microtubule motor protein, CENP-E (Zecevic et al. 1998). The kinesin-like tobacco proteins, as well as CENP-E, specifically accumulate in mitosis. The microtubule association and mitosis-specific accumulation suggest that NPK1 might be an upstream activator of MMK1 and/or MMK3 and NTF6.

Another regulator of MMK3 and NTF6 could be the plant homologue of the *Schizosaccharomyces pombe cdc7* gene, a critical regulator of cytokinesis. *cdc7* encodes a protein kinase similar to the *Saccharomyces cerevisiae* STE11, a signalling component upstream of a MAPK module (Sohrman et al. 1998). The Cdc7 protein is associated with one of the spindle pole bodies in anaphase,

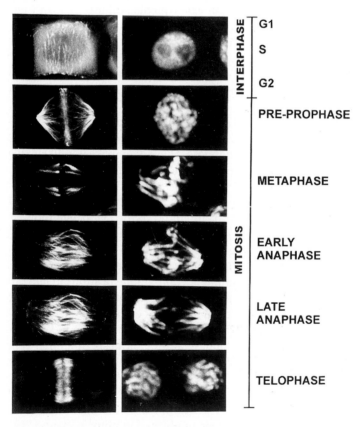

**Fig. 3.** Microtubule organisations during the plant cell cycle. Tubulin staining *left*, and DNA staining *right*. In interphase cells *1st row*, cortical microtubules, consisting of parallel microtubule arrays, which gather beneath the plasma membrane. In $G_2$ *2nd row*, the preprophase band forms, which is composed of bundled microtubules on the cell cortex at the position of the future division plane. In preprophase cells, another characteristic microtubule structure is the prophase spindle; which is composed of microtubules radiating from the division poles on the nuclear surface. In metaphase *3rd row*, the spindle typically connects kinetochores with broad division poles. In anaphase, *4th and 5th rows*, the anaphase spindle is present. In telophase *6th row*, the phragmoplast, composed of short interdigitating microtubule fibers is formed at the midplane of cell division

which ensures that the division plane is restricted to a single location. Plant cells do not have a well-defined microtubule-organising center, but the focusing of MMK3 and NTF6 from a diffuse distribution to a flat disk at the midplane of cell division might serve a similar function.

A dynamin-like protein, phragmoplastin, has been isolated from soybean and shown to be associated with cell plate formation (Gu and Verma 1996, 1997). Phosphorylation of dynamin by ERK2 inhibits the dynamin-microtubule interaction (Earnest et al. 1996). Phragmoplastin was found to appear first in the center of the forming cell plate and, as the cell plate grew outwards, it redistributed to the growing margins of the cell plate. Contrary to this, we have not found a redistribution of MMK3 from the center to the periphery, indicating that MMK3 is not associated with phragmoplastin during this process. Recently, embryo patterning genes in *Arabidopsis thaliana* such as *knolle* (Lukowitz et al. 1996) and *keule* (Assaad et al. 1996) have been identified as being involved in cytokinesis, and the syntaxin-related KNOLLE protein was localised at the cell plate (Lauber et al. 1997). Similarly to MMK3 in late telophase, when the phragmolast reached the lateral cortex of the cell, KNOLLE appeared to be present across the entire plane of division. Phragmoplast microtubules might be required to bring these proteins to the cell plate but apparently the microtubules are not required to keep KNOLLE and MMK3 at this location. Fusion of MMK3 with a green fluorescent protein will provide a visual tag to determine to what extent immunofluorescence labelling on fixed cells reflects the in vivo localisation of the MMK3 protein.

Which events of cytokinesis might be regulated by MMK3 and NTF6 is not known at present. They may include the construction of the phragmoplast by regulating microtubule stability or microtubule-based motor proteins (Chan et al. 1996; Liu et al. 1996; Mitsui et al. 1996; Asada et al. 1997; Bowser and Reddy 1997), vesicle transport along the phragmoplast by plus-end-directed motor proteins (Asada and Collins 1997), or the fusion of these vesicles at the cell plate (Verma and Gu 1996).

# 5
# Spatial Regulation of Cell Division and Growth are Dictated by Plant Specific Cytoskeletal Structures

Plants are multicellular organisms with cells in fixed positions. Thus, not only the regulation of the timing and the number of divisions but also the orientation of these divisions and the orientation of cell enlargement are critical to elaborate the plant body shape. These aspects of cell division are regulated by the interchange of five major plant-specific cytoskeletal structures: the cortical array, the radial cytoplasmic array, the preprophase band of cortical microtubules, the mitotic spindle and the phragmoplast (Figure 3). How these arrays interchange and how they are positioned are crucial during plant cell divisions and development. Here we focus our attention on the microtubule (MT) components of the plant cytoskeleton, but bear in mind that microfilaments are just as important as targets for signals.

## 5.1
## The Cortical Microtubules

Cortical microtubules encircle the interphase cell with parallel groups of microtubules with overlapping ends. Some are attached to the plasma membrane (Hush and Overall 1996). These MTs seem to be involved in channelling cellulose-synthesising complexes by guiding their movement across the face of the plasma membrane (Giddings and Staehelin 1991). Because these tough wall microfibrils resist turgor-driven cell expansion, the way in which they are wrapped around the cell determines the direction of cell growth. In an elongating cell, microfibrils are arranged transversely, whereas re-orientation of microtubules parallel to the existing growth axis causes the cell to swell laterally. Microtubules provide the frame for the cell even though they are surprisingly dynamic polymers, exhibiting a turnover rate of $t_{1/2}=67$ s, which is 3–4 times faster than interphase microtubules in mammalian cells (Hepler and Hush 1996). It is precisely this dynamic nature of microtubules which allows plants to respond rapidly to various stimuli by redirecting cell growth (Lloyd 1994). The list of stimuli is long and includes plant hormones, light, wound, electrical signals, mechanical stimuli, bacterial invasion etc. (Shibaoka 1994; Fisher and Schopfer 1997). How these signals are perceived and lead to the redirection of microtubule growth is not well understood. Observation of the re-orientation of microtubules in living cells after visualisation by incorporation of fluorescently labelled tubulin showed that no wholesale depolymerisation occurs. Rather, increased numbers of microtubules in a new alignment appear, and gradually replace the previous arrays, which could involve some selective stabilisation of microtubules in the new direction (Wymer and Lloyd 1996). Most of these stimuli have been shown to induce MAPKs as well as the release of $Ca^{2+}$ before the re-organisation of microtubules (Hirt 1997). That protein phosphorylation is an important regulatory component in the orientation of cortical microtubules was shown by the inhibition of MT re-orientation upon gibberellic acid treatments into transverse arrangements if protein kinases were inhibited with the drugs staurosporine and 6 DMAP (Mizuno 1994; Mayumi and Shibaoka 1996; Takesue and Shibaoka 1998). These drugs are known to inhibit MAPK pathways both in yeast (Toda et al. 1991) and in plant cells (Suzuki and Shinshi 1995). Selective destabilisation of microtubules in the previous direction and their stabilisation in a new direction might involve the phosphorylation/dephosphorylation of microtubule-associated proteins (MAPs). Several plant MAPs have been described and were shown to associate with cortical microtubules (Cyr and Palevitz 1989; Jiang and Sonobe 1993; Schellenbaum et al. 1993; Vantard et al. 1994; Chan et al. 1996). MAPK was shown to associate with microtubules and regulate microtubule stability in animal cells (Reszka et al. 1995; Morishima-Kawashima and Kosik 1996).

Phosphorylation of tau protein, a mediator of the interaction between microtubules and the plasma membrane, could provide another target for influencing cortical microtubules. Tau-like proteins have been identified in association with cortical microtubules (Nick et al. 1995). Tau is a substrate for MAPKs in animal cells (Drewes et al. 1992) and a tau-related protein was also phosphorylated in vitro in plant cell extracts (Vantard et al. 1991, 1994).

## 5.2
## The Preprophase Band

The division plane is marked by a band of cortical microtubules aligned with the nuclear equator which forms before mitosis in plant cells (Wick 1991). This preprophase band (PPB) is first recognised as a wide band of cortical MTs at the end of S-phase or in the early $G_2$ phase which is usually superseded by a narrower band during $G_2$ phase progression. The PPB could be formed by the condensation or "bunching up" of existing cortical microtubules or by selective stabilisation or polymerisation of microtubules in a new location. The orientation of cortical microtubules is critical for positioning the PPB. This was shown when pea root cells responded to a nearby wound by re-entering the division cycle and dividing in planes that are parallel to the cut surface. In this experiment, interphase microtubules were re-oriented into arrays parallel to the wound, and then formed PPBs in the new orientation (Hush et al., 1990). These wound-induced divisions can be further re-oriented into a new plane by re-wounding, providing that the second wound is imposed within a limited time period, indicating that the determination of the cell division plane is a stepwise process which can be reverted within a certain time interval (Venverloo 1990). Another signal which determines PPB position must come from the nucleus, because misplacement of the nucleus by centrifugation induced the formation of the PPB at the new location, provided that it happened before or at early stages of PPB formation (Murata and Wada 1991). The influence of the nucleus on PPB position could be provided by a gradient of phosphorylation determined by protein kinases and phosphatases located in the nucleus. It may be relevant that $G_2$-specific cyclins, possibly in association with CDKs, localise to the nucleus at this time (Mews et al. 1997).

Thus it appears that division site selection is a flexible event, which is gradually established throughout $G_2$ in the cortex by the culmination of signals originating both from the cell surface and from the nucleus. What these signals are and how they determine the position of the PPB is not yet known.

Protein phosphorylation regulates microtubule stability and preprophase band formation as shown by experiments using the protein kinase inhibitors K252-a and staurosporine. K252-a or staurosporine, when applied in the early $G_2$ phase, arrested the cell cycle before mitosis and inhibited the formation of the preprophase band, while treatment in late $G_2$ blocked the disassembly of the preprophase band (Katsuta and Shibaoka 1992). Conditions favouring protein phosphorylation, such as the presence of ATP, induce the disassembly of microtubules in in vitro permeabilised cells. This effect was inhibited by staurosporine, but not by K252-a, indicating the involvement of various protein kinases with distinct sensitivities to these drugs (Katsuta and Shibaoka 1992). Treatment of cells with K252-a and staurosporine inhibits the activation of protein kinases with properties similar to MAPKs (Suzuki and Shinshi 1995; Ádám et al. 1997).

CDK was found to co-localise with the PPB in maize root tip cells (Colasanti et al. 1993). Microinjection of active CDK into plant cells led to the depolymerisation of the preprophase band (Hush et al. 1996), while the inhibition of CDK activity with the drug roscovitine inhibited this process (Binarova et al. 1998b), indicating that CDK might regulate the interchange of cortical microtubules with the mitotic spindle. Given the overlap in substrate specificity between MAPK and CDK, it is still an open question as to which of these protein kinases performs this function in vivo (Verlhac et al. 1994). As the transcription and protein levels of MMK1 and NPK1 correspond to the timing of PPB formation, these kinases are plausible candidates to be involved in this process.

## 5.3
## The Mitotic Spindle

Conversion of the interphase microtubule array to a mitotic spindle is a dramatic example of cytoplasmic reorganisation. Two mutually exclusive models have been proposed to explain the mechanism of spindle assembly: the "search and capture " model suggests that dynamic microtubules are stabilised and thus captured by kinetochores on chromosomes (Mitchison 1988); while the "local stabilisation" model suggests that chromosomes create a change in the state of the cytoplasm around them which is more favourable to microtubule polymerisation (Waters and Salmon 1997). Moreover, spindle formation seems to be an intrinsic property of the microtubule network, because bipolar spindles could be formed even in the absence of chromosomes (Brunet et al. 1998). Higher plants lack centrosomes, which are spatial clues on the poles for bipolar spindle organisation in somatic animal cells. Dispersed microtubule organising centers were hypothesised to explain microtubule nucleation during spindle formation in acentriolar plant cells (Mazia 1987). Another suggestion is the cytoplasmic self-assembly of microtubules which might play a critical role in plant cells (Smirnova and Bajer 1998). The nuclear envelope, however, is able to promote the nucleation of microtubules in a cell cycle-specific manner (Lambert 1993; Vaughn and Harper 1998).

Both in animal and in plant cells, microtubules are stabilised by microtubule-associated proteins, (MAPs) some of which specifically interact with the mitotic spindle (Chan et al. 1996; Andersen and Karsenti 1997). In mitosis, the stabilising effect of MAPs on MTs is inactivated by phosphorylation as cells enter mitosis (Gotoh et al. 1991; Shiina et al. 1992; Ookata et al. 1995). Similarly, mitotic plant extracts enriched in MAPs result in the nucleation of shorter microtubules than interphase extracts (Stoppin et al. 1996). Beside MAPs, which stabilise microtubules, there are additional regulators of microtubule dynamics which actively increase microtubule depolymerisation. One such factor is stathmin/Op18, which inhibits microtubule polymerisation by interacting with tubulin (Belmont and Mitchison 1996). This ability of stathmin/Op18 to depolymerise microtubules is inactivated by phosphorylation on sites which are potential targets for Cdks and mitogen-activated protein kinases (Marklund et al. 1996). The protein phosphatase PP2A counteracts this effect by dephosphorylating stathmin/Op18 at these sites (Tournebize et al. 1997). Some of the protein kinase activity responsible for stathmin/Op18 phosphorylation is associated with mitotic chromatin (Andersen et al. 1997).

The finding of $\gamma$-tubulin, a tubulin isoform involved in microtubule nucleation and stabilisation, in association with kinetochores suggests that plant chromosomes might play a special role in nucleating and stabilising kinetochore fibers (Binarova et al. 1998a).

The step following the selective polymerisation of microtubules around mitotic chromatin is the organisation of these randomly oriented microtubule fibers into a bipolar mitotic spindle. The self-organisation of a bipolar spindle around chromatin-coated beads in a mitotic *Xenopus* extract in vitro demonstrated that this step requires neither centrosomes nor centromeres but that it is organised by plus- and minus-end-directed microtubule motors in these extracts (Heald et al. 1996). First, a multiheaded motor complex (such as the one which is formed from Eg5 molecules) crosslinks microtubules, and as the motor-complex moves toward the plus end, the parallel microtubules become orientated (Merdes and Cleveland 1997). Recently, a molecular motor similar to Eg5 has also been isolated from tobacco and possesses a consensus phosphorylation site for CDKs and MAPKs (Asada et al. 1997).

The final step in achieving a spindle is focusing the minus ends of the microtubules to the poles. This was shown to be performed by the minus end-directed molecular motor, dynein, in a complex with a non-motor protein, NuMA (Merdes and Cleveland 1997).

Similarly to microtubule stability, microtubule movements are also regulated by phosphorylation during mitosis (Verde et al. 1991). Eg5 is phosphorylated by CDK1 on a site conserved in yeasts, animals, and higher plants, and this phosphorylation is required for the association of Eg5 with the mitotic spindle (Blangy et al. 1995; Sawin and Mitchison 1995). However, the overlapping substrate specificities between MAPKs and CDKs must be taken into consideration.

Monoclonal antibodies have been developed which specifically recognise mitosis-specific phosphoepitopes. The MPM2 antibody stains the nucleus of interphase plant cells and the intensity of staining increases during $G_2$ and the formation of the preprophase band (Binarova et al. 1993, 1994). In prophase and metaphase the MPM2 antibody decorates the plant kinetochores (Binarova et al. 1994). The MPM2 phosphoepitope is phosphorylated by proline-directed protein kinases, including MAPKs (Kuang and Ashorn 1993). The MPM2 antibody was also shown to bind to phosphorylated MAPK and thereby inhibit its activity (Taagepera et al. 1994). Another mitosis-specific phosphoepitope is recognised by the 3F3/2 monoclonal antibody. This antibody recognises kinetochores of mitotic animal cells with missaligned chromosomes and detects phosphorylations involved in spindle assembly checkpoint (Gorbsky 1997; Nicklas 1997). The ERK1 and ERK2 MAPKs were suggested to create the 3F3/2 phosphoepitope on animal kinetochores (Shapiro et al. 1998). Specific localisation of protein kinases and protein phosphatases might create a gradient of phosphorylation, which could provide a mechanism regulating preferential microtubule growth around chromosomes.

CDK activity appears to be crucial in establishing the bipolar spindle, because treatment of cells with roscovitine inhibited spindle formation. Thus, chromosomes could not be aligned on the metaphase plate, but were trapped by a monopolar spindle in a circular configuration, with kinetochores pointing inward and chromosome arms outward (Binarova et al. 1998b). Roscovitine also inhibited plant MAPKs, but approximately ten times less efficiently. However, the roscovitine concentration which was required to obtain mitotic abnormalities already inhibits MAPKs in vitro. Thus, it is possible that some of these abnormalities could be attributed to MAPKs.

## 5.4
## The Phragmoplast

Rather than dividing up the cytoplasm by constriction from the plasma membrane as in animals (Glotzer 1997), plant cells construct a new cell plate from the inside (Staehelin and Hepler 1996). The phragmoplast appears between the daughter nuclei at late anaphase. It consists of microtubules and microfilaments which are organised as two parallel interdigitating strands in the plane of cell division. Golgi-derived vesicles and elements of the endoplasmic reticulum are transported along these cytoskeletal arrays. Later the vesicles fuse to form first a continuous tubular network and then gradually consolidate into a flattened disk in which callose is deposited (Samuels et al. 1995).

The phragmoplast MTs appear to arise from remnants of the mitotic spindle apparatus. In late anaphase the loosely splayed MTs in the interzone coalesce laterally into a centrally positioned cylindrical structure, which subsequently shortens in length and expands in girth until it finally reaches the side wall. The short phragmoplast MTs align perpendicular to the plane of the future cell plate with their plus ends overlapping. The MTs of the phragmoplast are also dynamic, exibiting rapid turnover ($t_{1/2}$=60 s). Thus, construction of the phragmoplast from the anaphase spindle could be achieved by selective stabilisation of MTs in the midzone, or by redistribution of microtubules by motor proteins (Asada et al. 1991). A kinesin motor, TPRK125, was isolated from phragmoplasts and shown to translocate microtubules. TPRK125 is structurally related to the Eg5 motor from animal cells and contains a consensus phosphorylation site for CDKs and MAPKs (Asada et al. 1997).

Cyclin-dependent kinases have been central in understanding the control of cell division. However, other protein kinases, including MAPKs, placed in parallel, upstream, or downstream of CDKs, also regulate cell divisions. In plants we are beginning to isolate the components, but little is known about how they are connected, which is crucial in understanding how different stimuli influence cell divisions and thereby meristem function and plant development.

*Acknowledgments.* The authors wish to thank Cathal Wilson and John Beyerly for critical reading of the manuscript and Ota Blahousek for photography. Related work in our laboratories has been promoted by grants GA AVCR 5020803/1998, Aktion Österreich-Tchechische Republik and the TMR project EU-CYTO FMRX-CT98.

# References

Ádám AL, Pike S, Hoyos ME, Stone JM, Walker JC, Novacky A (1997) Rapid and transient activation of a myelin basic protein kinase in tobacco leaves treated with harpin from *Erwinia amylovora*. Plant Physiol 115:853–861

Andersen SSL, Karsenti E (1997) XMAP310: a *Xenopus* rescue-promoting factor localised to the mitotic spindle. J Cell Biol 139:975–983

Andersen SSL, Ashford AJ, Tournebize R, Gavet O, Sobel A, Hyman AA, Karsenti E (1997) Mitotic chromatin regulates phosphorylation of stathmin/Op18. Nature 389:640–643

Asada T, Collings D (1997) Molecular motors in higher plants. Trends Plant Sci 2:29–37

Asada T, Sonobe S, Shibaoka H (1991) Microtubule translocation in the cytokinetic apparatus of cultured tobacco cells. Nature 350:238–241

Asada T, Kuriyama R, Schibaoka H (1997) TKRP125, a kinesin-related protein involved in the centrosome-independent organisation of the cytokinetic apparatus in tobacco By-2 cells. J Cell Sci 110:179–189

Assaad FF, Mayer U, Wanner G, Jürgens G (1996) The *KEULE* gene is involved in cytokinesis in *Arabidopsis*. Mol Gen Genet 253:267–277

Banno H, Hirano K, Nakamura T, Irie K, Nomoto S, Matsumoto K, Machida Y (1993) *NPK1*, a tobacco gene that encodes a protein with a domain homologous to yeast BCK1, STE11, and BY2 protein kinases. Mol Cell Biol 13:4745–4752

Becraft PW (1998) Receptor kinases in plant development. Trends Plant Sci 3:384–388

Belmont LD, Mitchison TJ (1996) Identification of a protein that interacts with tubulin dimers and increases the catastrophe rate of microtubules. Cell 84:623–631

Binarova P, Cihalikova J, Dolezel J (1993) Localization of MPM-2 recognized phosphoproteins and tubulin during cell cycle progression in synchronized *Vicia faba* root meristem cells. Cell Biol Int 9:847–856

Binarova P, Rennie P, Fowke L (1994) Probing microtubule organizing centers with MPM-2 in dividing cells of higher plants using immunofluorescence and immunogold technique. Protoplasma 180:106–117

Binarova P, Hause B, Dolezel J, Draber P (1998a) Association of γ-tubulin with kinetochore/centromeric region of plant chromosomes. Plant J 14:751–757

Binarova P, Dolezel J, Draber P, Heberle-Bors E, Strnad M, Bögre L. (1998b). Treatment of *Vicia faba* root tip cells with specific inhibitors to cyclin-dependent kinases leads to abnormal spindle formation. Plant J 16:697–707

Blangy A, Lane HA, de'Hérin P, Harper M, Kress M, Nigg EA (1995) Phosphorylation by p34$^{cdc2}$ regulates spindle association of human Eg5, a kinesin-related motor essential for bipolar spindle formation in vivo. Cell 83:1159–1169

Bögre L, Jonak C, Mink M, Meskiene I, Traas J, Ha DCH, Swoboda I, Plank C, Wagner E, Heberle-Bors E, Hirt H (1996) Developmental and cell cycle regulation of alfalfa *nucMs1*, a plant homolog of the yeast Nsr1 and mammalian nucleolin. Plant Cell 8:417–428

Bögre L, Calderini O, Binarova P, Mattauch M, Till S, Kiegerl S, Jonak C, Pollaschek C, Baker P, Huskisson NS, Hirt H, Heberle-Bors E (1999) A MAP kinase is activated late in mitosis and becomes localised to the plane of cell division. Plant Cell 11:101–103

Bottazzi ME, Assoian RK (1997) The extracellular matrix and mitogenic growth factors control G1 phase cyclins and cyclin-dependent kinase inhibitors. Trends Cell Biol 7:348–352

Bowser J, Reddy ASN (1997) Localization of a kinesin-like calmodulin-binding protein in dividing cells of *Arabidopsis* and tobacco. Plant J 12:1429–1437

Brunet S, Polanski Z, Verlac M-H, Kubiak JZ, Maro B (1998) Bipolar meiotic spindle formation without chromatin. Curr Biol 8:1231–1234

Calderini O, Bögre L, Vicente O, Binarova P, Heberle-Bors E, Wilson C (1998) A cell cycle regulated MAP kinase with a possible role in cytokinesis in tobacco cells. J Cell Sci 111:3091–3100

Campbell MS, Gorbsky GJ (1995) Microinjection of mitotic cells with the 3F3/2 anti-phosphoepitope antibody delays the onset of anaphase. J Cell Sci 129:1195–1204

Chan J, Rutten T, Lloyd C (1996) Isolation of microtubule-associated proteins from carrot cytoskeletons: a 120 kDa MAP decorates all four microtubule arrays and the nucleus. Plant J 10:251–259

Chiri S, De Nadai C, Ciapa B (1998) Evidence for MAP kinase activation during mitotic division. J Cell Sci 111:2519–2527

Colasanti J, Cho SO, Wick S, Sundaresan V (1993) Localization of the functional p34$^{cdc2}$ homolog of maize in root tip and stomatal complex cells: association with predicted division sites. Plant Cell 5:1101–1111

Cyr RJ, Palevitz BA (1989) Microtubule-binding proteins from carrot. I. Initial characterisation and microtubule bundling. Planta 177:245–260

Dahl M, Meskiene I, Bögre L, Ha DTC, Swoboda I, Hubmann R, Hirt H, Heberle-Bors E (1995) The D-type alfalfa cyclin gene cycMs4 complements G$_1$ cyclin-deficient yeast and is induced in the G$_1$ phase of the cell cycle. Plant Cell 7:1847–1857

Drewes G, Lichtenberg-Kraag B, Doring F, Mandelkow EM, Biernat J, Goris J, Doree M, Mandelkow E (1992) Mitogen-activated protein (MAP) kinase transforms tau protein into an Alzheimer-like state. EMBO J 11:2131–2138

Earnest S, Khokhlatchev A, Albanesi JP, Barylko B (1996) Phosphorylation of dynamin by ERK2 inhibits the dynamin-microtubule interaction. FEBS Lett 396:62–66

Fisher K, Schopfer P (1997) Interaction of auxin, light, and mechanical stress in orienting microtubules in relation to tropic curvature in the epidermis of maize coleoptiles. Protoplasma 196:108–116

Fuerst RUA, Soni R, Murray JAH, Lindsey K (1996) Modulation of cyclin transcript levels in cultured cells of *Arabidopsis thaliana*. Plant Physiol 112:1023–1033

Galaktionov K, Jessus C, Beach D (1995) Raf1 interaction with Cdc25 phosphatase ties mitogenic signal transduction to cell cycle activation. Genes Dev 9:1046–1058

Giddings TH, Staehelin LA (1991) Microtubule-mediated control of microfibril deposition: a reexamination of the hypothesis. In:Lloyd CW(ed) The cytoskeletal basis of plant growth and form. Academic Press, London, pp 85–99

Glotzer M (1997) The mechanism and control of cytokinesis. Curr Opin Cell Biol 9:15–823
Gorbsky GJ (1997) Cell cycle checkpoints: arresting progress in mitosis. BioEssays 19:193–197
Gotoh Y, Nishida E, Matsuda S, Shiina N, Kosako H, Shiokava K, Akiyama T, Ohta K, Sakai H
    (1991) In vitro effects on microtubule dynamics of purified *Xenopus* M phase-activated MAP
    kinase. Nature 349:251–254
Gu X, Verma DPS (1996) Phragmoplastin, a dynamin-like protein associated with cell plate for-
    mation in plants. EMBO J 15:695–704
Gu X, Verma DPS (1997) Dynamics of phragmoplastin in living cells during cell plate formation
    and uncoupling of cell elongation from the plane of cell division. Plant Cell 9:157–169
Heald R, Tournebize R, Blank T, Sandaltzopoulos R, Becker P, Hyman A, Karsenti E (1996) Self-
    organisation of microtubules into bipolar spindles around artificial chromosomes in *Xenopus*
    egg extracts. Nature 382:420–425
Heider H, Hug C, Lucocq JM (1994) A 40 kDa myelin basic protein kinase, distinct from Erk1 and
    Erk2, is activated in mitotic HeLa cells. Eur J Biochem 219:513–520
Hepler PK, Hush JM (1996) Behavior of microtubules in living plant cells. Plant Physiol 112:
    455–461
Hirt H (1997) Multiple roles of MAP kinases in plant signal transduction. Trends Plant Sci 2:
    11–15
Hush J, Wu L, John PC, Hepler LH, Hepler PK (1996) Plant mitosis promoting factor disassem-
    bles the microtubule preprophase band and accelerates prophase progression in *Tradescan-
    tia*. Cell Biol Int 20:275–287
Hush JM, Overall RL (1996) Cortical microtubule reorientation in higher plants: dynamics and
    regulation. J Microsc 181:129–139
Hush JM, Hawes CR, Overall RL (1990) Interphase microtubule reorientation predicts a new cell
    polarity in wounded pea roots. J Cell Sci 96:47–61
Huttly AK, Phillips AL (1995) Gibberellin-regulated expression in oat aleurone cells of two kinas-
    es that show homology to MAP kinase and ribosomal protein kinase. Plant Mol Biol 27:
    1043–1052
Jiang CJ, Sonobe S (1993) Identification and preliminary characterisation of a 65 kDa higher
    plant microtubule-associated protein. J Cell Sci 105:891–901
Jonak C, Páy A, Bögre L, Hirt H, Heberle-Bors E (1993) The plant homologue of MAP kinase is
    expressed in a cell cycle-dependent and organ-specific manner. Plant J 3:611–617
Jones HD, Smith SJ, Desikan R, Plakidou-Dymock S, Lovegrove A, Hooley R (1998) Heterotri-
    meric G proteins are implicated in gibberellin induction of α-amylase gene expression in wild
    oat aleurone. Plant Cell 10:245–253
Kakimoto T (1998) Cytokinin signaling. Curr Opin Plant Biol 1:399–403
Katsuta J, Shiboka H (!992) Inhibition by kinase inhibitors of the development and the disap-
    pearance of the preprophase band of microtubules in tobacco BY-2 cells. J Cell Sci 103:
    397–405
Kennelly PJ, Krebs EG (1991) Consensus sequences as substrate specificity determinants for pro-
    tein kinases and protein phosphatases. J Biol Chem 266:15555–15558
Knetsch MLW, Wang M, Snaar-Jagalska BE, Heimovaara-Dijkstra S (1996) Abscisic acid induces
    mitogen-activated protein kinase activation in barley aleurone protoplasts. Plant Cell 8:
    1061–1067
Kovtun Y, Chiu W-L, Zeng W, Sheen J (1998) Suppression of auxin signal transduction by a
    MAPK cascade in higher plants. Nature 395:716–720
Kuang J, Ashorn CL (1993) At least two kinases phosphorylate the MPM-2 epitope during *Xeno-
    pus* oocyte maturation. J Cell Biol 123:859–868
Laird AD, Taylor SJ, Oberst M, Shalloway D (1995) Raf-1 is activated during mitosis. J Biol Chem
    270:26742–26745
Lambert AM (1993) Microtubule-organizing centres in higher plants. Curr Opin Cell Biol 5:
    116–122

Lauber MH, Waizenegger I, Steinmann T, Schwarz H, Mayer U, Hwang I, Lukowitz W, Jürgens G (1997) The *Arabidopsis* KNOLLE protein is a cytokinesis-specific syntaxin. J Cell Biol 139: 1485–1493

Lavoie JN, L'Allemain G, Brunet A, Müller R, Pouyssegur J (1996) Cyclin D1 expression is regulated positively by the p42/44$^{MAPK}$ and negatively by the p38/HOG$^{MAPK}$ pathway. J Biol Chem 271:20608–20616

Liao H, Li G, Yen TJ (1994) Mitotic regulation of microtubule crosslinking activity of CENP-E kinetochore protein. Science 265:394–398

Liu B, Cyr RJ, Palevitz BA (1996) A kinesin-like protein, KatAp in the cells of *Arabidopsis* and other plants. Plant Cell 8:119–132

Liu SH, Lee HH, Chen JJ, Chuang CF, Ng SY (1994) Serum response element-regulated transcription in the cell cycle: possible correletion with microtubule reorganisation. Cell Growth Differ 5:447–455

Lloyd CW (1994) Why should stationary plant cells have such dynamic microtubules? Mol Biol Cell 5:1277–1280

Lukowitz W, Mayer U, Jürgens G (1996) Cytokinesis in the *Arabidopsis* embryo involves the syntaxin-related KNOLLE gene product. Cell 84:61–71

Machida Y, Nakashima M, Morikiyo K, Banno H, Ishikawa M, Soyano T, Nishihama R (1998) MAPKKK-related protein kinase NPK1: regulation of the M phase of plant cell cycle. J Plant Res 111:243–246

Marklund U, Larsson N, Gradin MH, Brattsand G, Gullberg M (1996) Oncoprotein 18 is a phosphorylation-responsive regulator of microtubule dynamics. EMBO J 15:5290–5298

Marshall CJ (1996) Ras effectors. Curr Opin Cell Biol 8:197–204

Mayumi K, Shibaoka H (1996) The cyclic reorientation of cortical microtubules on walls with a crossed polylamellate structure: effects of plant hormones and an inhibitor of protein kinases on the progression of the cycle. Protoplasma 195:112–122

Mazia D (1987) The chromosome cycle and the centrosome cycle in the mitotic cycle. Int Rev Cytol 100:49–92

Merdes A, Cleveland DW (1997) Pathways of spindle pole formation: different mechanisms; conserved components. J Cell Biol 138:953–956

Meskiene I, Bögre L, Glaser W, Balogh J, Brandstötter M, Zwerger K, Ammerer G, Hirt H (1998) MP2C, a plant protein phosphatase 2C, functions as a negative regulator of mitogen-activated protein kinase pathways in yeast and plants. Proc Natl Acad Sci USA 95:1938–1943

Mews M, Sek, FJ, Moore R, Volkmann D, Gunning BES, John PCL (1997) Mitotic cyclin distribution during maize cell division: implications for the sequence diversity and function of cyclins in plants. Protoplasma 200:128–145

Meyerowitz EM (1997) Genetic control of cell division patterns in developing plants. Cell 88: 299–308

Minshull J, Sun H, Tonks NK, Murray AW (1994) A MAP kinase-dependent spindle assembly checkpoint in *Xenopus* egg extracts. Cell 79:475–486

Mitchison TJ (1988) Microtubule dynamics and kinetochore function in mitosis. Annu Rev Cell Biol 4:527–49

Mitsui H, Hasezawa S, Nagata T, Takahashi H (1996) Cell cycle-dependent accumulation of a kinesin-like protein, KatB/C, in sychronized tobacco BY-2 cells. Plant Mol Biol 30:177–181

Mizoguchi T, GotohY, Nishida E, Yamaguchi-Shinozaki K, Hayashida N, Iwasaki T, Kamada H, Shinozaki K (1994) Characterization of two cDNAs that encode MAP kinase homologues in *Arabidopsis thaliana* and analysis of the possible role of auxin in activating such kinase activities in cultured cells. Plant J 5:111–122

Mizuno K (1994) Inhibition of giberellin-induced elongation, reorientation of cortical microtubules and change of isoform of tubulin in epicotyl segments of azuki bean by protein kinase inhibitors. Plant Cell Physiol 35:1149–1157

Morishima-Kawashima M, Kosik KS (1996) The pool of MAP kinase associated with microtubules is small but constitutively active. Mol Biol Cell 7:893–905

Murata T, Wada M (1991) Effects of centrifugation on preprophase-band formation in *Adiantum* protonemata. Planta 183:391–398

Murray JAH (1997) The retinoblastoma protein is in plants. Trends Plant Sci 2:82–84

Nick P, Lambert AM, Vantard M (1995) A microtubule-associated protein in maize is expressed during phytochrome-induced cell elongation. Plant J 8:835–844

Nicklas RB (1997) How cells get the right chromosome. Science 275:632–637

Ookata K, Hisanaga S, Bulinski JC, Murofushi H, Aizawa H, Itoh TJ, Hotani H, Okumura E, Tachibana K, Kishimoto T (1995) Cyclin B interaction with microtubule-associated protein 4 (MAP4) targets p34cdc2 kinase to microtubules and is a potential regulator of M-phase microtubule dynamics. J Cell Biol 128:849–862

Palmer A, Gavin AC, Nebrada RA (1998) A link between MAP kinase and p34$^{cdc2}$/cyclin B during oocyte maturation: p90$^{Rsk}$ phosphorylates and inactivates the p34$^{cdc2}$ inhibitory kinase Myt1. EMBO J 17:5037–5047

Patham NI, Ashendel CL, Geahlen, RL, Harrison ML (1996) Activation of T cell Raf-1 at mitosis requires the protein-tyrosine kinase Lck. J Biol Chem 271:30315–30317

Peterson RT, Schreiber SL (1998) Connecting mitogens and the ribosome. Curr Biol 8:248–250

Philipova R, Whitaker M (1998) MAP kinase activity increases during mitosis in early sea urchin embryos. J Cell Sci 111:2497–2505

Plakidou-Dymock S, Dymock D, Hookley R (1998) A higher plant seven transmembrane receptor that influences sensitivity to cytokinins. Curr Biol 8:315–324

Reszka AA, Seger R, Diltz CD, Krebs EG, Fischer EH (1995) Association of mitogen-activated protein kinase with the microtubule cytoskeleton. Proc Natl Acad Sci USA 92:8881–8885

Robinson MJ, Cobb MH (1997) Mitogen-activated protein kinase pathways. Curr Opin Cell Biol 9:180–186

Roche S, Fumagalli S, Courtneidge SA (1995) Requirement for Src family protein tyrosine kinases in G2 for fibroblast cell division. Science 269:1567–1569

Rommel C, Hafen E (1998) Ras – a versatile cellular switch. Curr Biol 8:412–418

Sagata N (1997) What does Mos do in oocytes and somatic cells? BioEssays 19:13–21

Samuels AL, Giddings TH, Staehelin LA (1995) Cytokinesis in tobacco BY-2 and root tip cells: a new model of cell plate formation in higher plants. J Cell Biol 130:1345–1357

Sawin KE, Mitchison TJ (1995) Mutations in the kinesin-like protein Eg5 disrupting localisation to the mitotic spindle. Proc Natl Acad Sci USA 92:4289–4293

Schellenbaum P, Vantard M, Peter, C, Fellous A, Lambert AM (1993) Coassembly properties of higher plant microtubule-associated proteins with purified brain and plant tubulins. Plant J 3:253–260

Shapiro PS, Vaisberg E, Hunt AJ, Tolwinski NS, Whalen AM, McIntosh R, Ahn NG (1998) Activation of the MKK/ERK pathway during somatic cell mitosis: direct interactions of active ERK with kinetochores and regulation of the mitotic 3F3/2 phosphoantigen. J Cell Biol 142:1533–1545

Shiina N, Moriguchi T, Ohta K, Gotoh Y, Nishida E (1992) Regulation of a major microtubule-associated protein by MPF and MAP kinase. EMBO J 11:3977–3984

Smirnova E, Bajer AS (1998) Early stages of spindle formation and independence of chromosome and microtubule cycles in *Haemantus* endosperm. Cell Motil Cytoskel 40:22–37

Sohrmann M, Schmidt S, Hagan I, Simanis V (1998) Asymmetric segregation on spindle poles of the *Schizosaccharomyces pombe* septum-inducing protein kinase Cdc7p. Genes Dev 12:84–94

Solano R, Ecker JR (1998) Ethylene gas: perception, signaling and response. Curr Opin Plant Biol 1:393–398

Staehelin LA, Hepler PK (1996) Cytokinesis in higher plants. Cell 84:821-824

Stoppin V, Lambert AM, Vantard M (1996) Plant microtubule-associated proteins (MAPs) affect microtubule nucleation and growth at the plant nuclei and mammalian centrosomes. Eur J Cell Biol 69:11–23

Suzuki K, Shinshi H (1995) Transient activation and tyrosine phosphorylation of a protein kinase in tobacco cells treated with fungal elicitor. Plant Cell 7:639–647

Taagepera S, Dent P, Her JH, Sturgill TW, Gorbsky GJ (1994) The MPM-2 antibody inhibits mitogen-activated protein kinase activity by binding to an epitope containing phosphothreonine-183. Mol Biol Cell 5:1243–1251

Takenaka K, Gotoh Y, Nishida E (1997) MAP kinase is required for the spindle assembly checkpoint but is dispensable for the normal M phase entry and exit in *Xenopus* egg cell cycle extracts. J Cell Biol 136:1091–1097

Takenaka K, Moriguchi T, Nishida E (1998) Activation of the protein kinase p38 in the spindle assembly checkpoint and mitotic arrest. Science 280:599–602

Takesue K, Shibaoka H (1998) The cyclic reorientation of cortical microtubules in epidermal cells of azuki bean epicotyls: the role of actin filaments in the progression of the cycle. Planta 205: 539–546

Tamemoto H, Kadowaki T, Tobe K, Ueki K, Izumi T, Chatan Y, Kohno M, Kasuga M, Yazaki Y, Akanuma Y (1992) Biphasic activation of two mitogen-activated protein kinases during the cell cycle in mammalian cells. J Biol Chem 267:20293–20297

Taylor SJ, Shalloway D (1996) Cell cycle-dependent activation of Ras. Curr Biol 6:1621–1627

Tena G, Renaudin J-P (1998) Cytosolic acidification but not auxin at physiological concentration is an activator of MAP kinases in tobacco cells. Plant J 16:173–182

Thomas G, Hall MN (1997) TOR signalling and control of cell growth. Curr Opin Cell Biol 9: 782–787

Toda T, Shimanuki M, Yanagida M (1991) Fission yeast genes that confer resistance to staurosporine encode an AP-1-like transcription factor and a protein kinase related to the mammalian ERK1/MAP2 and budding yeast FUS3 and KSS1 kinases. Genes Dev 5:60–73

Tournebize R, Andersen SSL, Verde F, Dorée M, Karsenti E, Hyman AA (1997) Distinct roles of PP1 and PP2A-like phosphatases in control of microtubule dynamics during mitosis. EMBO J 16:5537–5549

Treisman R (1996) Regulation of transcription by MAP kinase cascades. Curr Opin Cell Biol 8: 205–215

Vantard M, Schellenbaum P, Fellous A, Lambert AM (1991) Characterisation of maize microtubule-associated proteins, one of which is immunologically related to tau. Biochemistry 30: 9334–9340

Vantard M, Peter C, Fellous A, Schellenbaum P, Lambert AM (1994) Characterisation of a 100 kDa heat-stable microtubule-associated protein from higher plants. Eur J Biochem 220: 847–853

Van't Hof (1974) Control of the cell cycle in higher plants. In: Pallida G.M, Cameron IL, Zimmerman A (eds) Cell Cycle Controls. Academic Press, New York, pp 77–85

Vaughn KC, Harper JD (1998) Microtubule-organising centers and nucleating sites in land plants. Int Rev Cytol 181:75–149

Venverloo CJ (1990) Regulation of the plane of cell division in vacuolated cells. II. Wound-induced changes. Protoplasma 155:85–94

Verde F, Berrez J-M, Antony C, Karsenti E (1991) Taxol-induced microtubule asters in mitotic extracts of *Xenopus* eggs: requirement for phosphorylated factors and cytoplasmic dynein. J Cell Biol 112:1177–1187

Verlhac MH, Kubiak JZ, Clarke HJ, Maro B (1994) Microtubule and chromatin behavior follow MAP kinase activity but not MPF activity during meiosis in mouse oocytes. Development 120:1017–1025

Verma DPS, Gu X (1996) Vesicle dynamics during cell plate formation in plants. Trends Plant Sci 1:145–149

Walker L, Estelle M (1998) Molecular mechanisms of auxin action. Curr Opin Plant Biol 1: 434–439

Wang H, Qi Q, Schorr P, Cutler AJ, Crosby WL, Fowke LC (1998) ICK1, a cyclin-dependent protein kinase inhibitor from *Arabidopsis thaliana* interacts with both Cdc2a and CycD3 and its expression is induced by abscisic acid. Plant J 15:501–510

Wang XM, Zhai Y, Ferrell JE (1997) A role for mitogen-activated protein kinase in the spindle assembly checkpoint in XTC cells. J Cell Biol 137:433–443

Waters JC, Salmon ED (1997) Pathways of spindle assembly. Curr Opin Cell Biol 9:37–43

Wick SM (1991) Spatial aspects of cytokinesis in plant cells. Curr Opin Cell Biol 3:253–260

Wilson C, Voronin V, Touraev A, Vicente O, Heberle-Bors E (1997) A developmentally regulated MAP kinase activated by hydration in tobacco pollen. Plant Cell 9:2093–2100

Wilson C, Pfosser M, Jonak C, Hirt H, Heberle-Bors E, Vicente O (1998) Evidence for the activation of a MAP kinase upon phosphate-induced cell cycle re-entry in tobacco cells. Physiol Plant 102:532–538

Wolniak SM, Larsen P (1995) The timing of protein kinase activation events in the cascade that regulates mitotic progression in *Tradescantia* stamen hair cells. Plant Cell 7:431–445

Wood KW, Sakowicz R, Goldstein LS, Cleveland DW (1997) CENP-E is a plus end-directed kinetochore motor required for metaphase chromosome alignement. Cell 91:357–366

Wymer C, Lloyd C (1996) Dynamic microtubules: implications for cell wall patterns. Trends Plant Sci 1:222–228

Zecevic M, Catling AD, Eblen ST, Renzi L, Hittle JC, Yen TJ, Gorbsky GJ, Weber MJ (1998) Active MAP kinase in mitosis: localisation at kinetochores and association with motor protein CENP-E. J Cell Biol 142:1547–1558

Zhang S, Klessig DF (1998) The tobacco wounding-activated protein kinase is encoded by SIPK. Proc Natl Acad Sci USA 95:7433–7438

Zhang SH, Broom MA, Lawton MA, Hunter T, Lamb CJ (1994) atpk1, a novel ribosomal protein kinase gene from *Arabidopsis* II. Functional and biochemical analysis of the encoded protein. J Biol Chem 269:17593–17599

Zhu X, Assoian RK (1995) Integrin-dependent activation of MAP kinase: a link to shape-dependent cell proliferation. Mol Biol Cell 6:273–282

# The MAP Kinase Cascade That Includes MAPKKK-Related Protein Kinase NPK1 Controls a Mitotic Process in Plant Cells

Ryuichi Nishihama and Yasunori Machida[1]

**Summary:** The tobacco *NPK1* cDNA was the first-isolated plant cDNA for a homolog of mitogen-activated protein kinase kinase kinases (MAPKKKs). The kinase domain of the NPK1 protein can replace the functions of MAPKKKs in yeasts, while the amino acid sequence of the kinase-unrelated region does not have any homology to those of MAPKKKs from other organisms. Transcription of the *NPK1* gene takes place in meristematic tissues or immature organs in a tobacco plant. During a tobacco cell cycle, transcriptional and translational products of *NPK1* are present from S to M phase and decrease after the M phase. Expression of the *NACK1* gene, which is predicted to encode a novel kinesin-like microtubule-based motor protein capable of activating NPK1, is specific to M phase, suggesting that activation of NPK1 occurs in M phase. Characterization of cDNAs for a MAPKK and a MAPK which can act downstream of NPK1 makes a proposition that the MAP kinase pathway involving NPK1 regulates a mitotic process associated with microtubules.

## 1
## Isolation of the *NPK1* cDNA

The proliferation of plant cells exhibits several unique features. The most characteristic features of this process are observed during cytokinesis and elongation of plant cells, which are tightly coupled with the formation of new cell walls inside and outside the cells, respectively. The formation of new cell walls is controlled by specially organized microtubules (Giddings and Staehelin 1991; Williamson 1991; Staehelin and Hepler 1996). These observations suggest that proliferation of plant cells may involve mechanisms characteristic of plants as well as those that function generally in eukaryotic cells. However, an understanding of the molecular mechanisms that regulate these processes in plant cells is almost completely lacking, although architectures and some com-

---

[1] Laboratory of Plant Development, Division of Biological Science, Graduate School of Science, Nagoya University, Chikusa-ku, Nagoya 464-8602, Japan

Results and Problems in Cell Differentiation, Vol. 27
Hirt (Ed.): MAP Kinases in Plant Signal Transduction
© Springer-Verlag Berlin Heidelberg 2000

ponents of plant-specific microtubule organizations have been reported (Asa-da et al. 1996; Gu and Verma 1996; Staehelin and Hepler 1996; Lauber et al. 1997), and a number of plant homologs of division-controlling genes of yeast and animal cells such as cyclin genes (Doerner 1994), cyclin-dependent protein kinase (CDK) genes (Doerner 1994) and *Rb* genes (Grafi et al. 1996; Xie et al. 1996; Ach et al. 1997) have been described.

At the beginning of the 1990s, in order to understand the intracellular events that occur in association with the cell proliferation of plant cells, we attempted to isolate tobacco homologs of cyclic AMP-dependent protein kinase (PKA) and $Ca^{2+}$-activated, phospholipid-dependent protein kinase (PKC) because they were proposed to play central roles in regulation of cellular functions, including cell proliferation, and they are ubiquitous in both animals and yeasts (Levin et al. 1990; Nishizuka 1992; Toda et al. 1993). To clone segments of cDNAs for such homologs, polymerase chain reaction was performed using cDNAs as templates which were prepared from messenger RNAs from the tobacco cultured cell line BY-2, which can provide a number of advantages for analysis of cell division (Banno et al. 1993b). Although DNA clones that appeared to encode homologs of either PKA or PKC were not isolated, several cDNAs that potentially encoded protein kinases were cloned in the course of this trial.

Since we were also interested in protein kinases that might be present and/or activated in dividing cells, Northern blot analysis was then carried out. The results showed that transcripts which were hybridized with one of the cDNAs accumulated in BY-2 cells during the logarithmic phase but not during the stationary phase, and that, in a tobacco plant, they were present in meristematic tissues and developmentally young organs but not in mature leaves (Banno et al. 1993a). These results suggested functions of this gene in proliferating cells (see below).

Although no protein that was homologous in terms of amino acid sequences to the predicted protein was found in databases when we had completed analysis of the nucleotide sequence of the cDNA, we meanwhile noticed that the deduced amino acid sequence was closely related to those of the protein kinase encoded by the *STE11, Byr2* and *BCK1* genes of the budding yeast which belong to the mitogen-activated protein kinase (MAPK) kinase kinase (MAPKKK) family: the MAPK cascade had just been established in yeasts. The observations described above allowed us to further characterize the predicted protein kinase, which was designated <u>N</u>icotiana <u>P</u>rotein <u>K</u>inase1 (NPK1; Banno et al. 1993a).

# 2
# Identification of NPK1 as a MAPKKK

The catalytic domain of NPK1 functions as a MAPKKK in yeast. The *BCK1* gene of *Saccharomyces cerevisiae* encodes a MAPKKK which acts downstream of PKC1, a budding yeast homolog of PKC (Levin et al. 1990; Lee and Levin 1992). The pathway mediated by BCK1 includes MKK1/2 as MAPKKs (Irie et al. 1993) and MPK1 (SLT2) as a MAPK (Torres et al. 1991; Lee et al. 1993). A mutation in any one of these genes for these kinases leads to a temperature-sensitive cell lysis defect in a conventional medium. Thus, this pathway is thought to mediate the signal which is generated in response to weakness in the cell wall under high temperature stresses (Kamada et al. 1995). Overexpression of a truncated NPK1 cDNA (NPK1ΔC cDNA), which encodes only the catalytic domain (see Fig. 1), complements the *bck1* mutant, demonstrating that NPK1 can act as a MAPKKK (Banno et al. 1993a). Although overexpression of the full-length NPK1 cDNA also replaces the growth function of *BCK1*, the degree of complementation by NPK1ΔC cDNA is greater than that by the full-length NPK1 cDNA. NPK1ΔC is also able to suppress the *pkc1* mutant whereas NPK1 is not. These data suggest that the NPK1ΔC protein has an activity which is higher than that of the full-length NPK1 protein, implying that the kinase-unrelated region of NPK1 negatively regulates its kinase activity.

The data obtained in the earlier study showed that NPK1 was specific to the *BCK1* pathway because it did not complement the *ste11* mutation (Banno et al. 1993a). Recent analysis with an activator of NPK1 (see below), however, showed that complementation of yeast mutants by NPK1 was not specific to the BCK1 pathway. Our unpublished data indicated that co-expression of the *NPK1* cDNA and a cDNA for the activator replaced functions of STE11 and SSK2/SSK22, MAPKKKs in other MAPK cascades of the budding yeast that act in the mating pheromone-response pathway and the osmosensing pathway, respectively (Nishihama et al. 1995). In the absence of the activator, such functional replacements were not observed. The mating pheromones activate a MAP kinase cascade that is composed of STE11, STE7 (MAPKK), and FUS3/KSS1 (MAPKs), resulting in the induction of trancription of the *FUS1* gene. The co-expression of the *NPK1* cDNA and its activator cDNA induced *FUS1* gene expression in the *ste11* mutant without exposure to the mating pheromones. This result suggests that NPK1 can activate the downstream MAPKK STE7 in the presence of its activator in yeast cells. The result of an experiment with the *ssk2/ssk22/sho1* triple mutant of yeast also suggested that NPK1 can activate PBS2, a MAPKK which acts downstream of SSK2/SSK22 in the pathway which transmits a signal generated by the high-osmolarity environment. Thus, NPK1 can act as a MAPKKK with a wide-ranging specificity in yeast cells.

# 3
# Structural Features of NPK1
# and Its *Arabidopsis* Homologs, ANP1/2/3

The kinase catalytic domain of NPK1 is located on the amino-terminal half, unlike the STE11-type MAPKKs which have them at the carboxyl termini

**A**

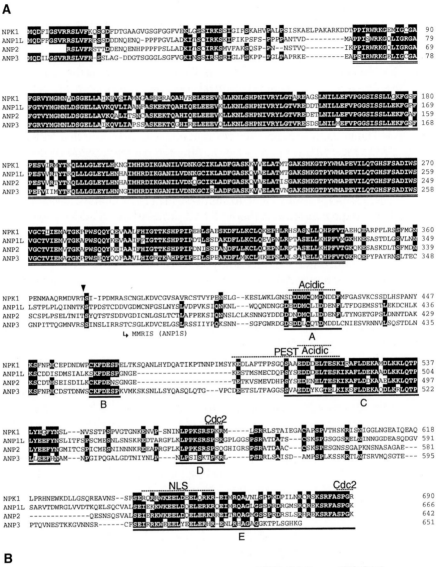

**B**

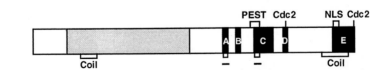

(Fig. 1). NPK1 has the kinase-unrelated region at the carboxyl terminus, the predicted amino acid sequence of which exhibits no homology in terms of amino acid sequence to any protein from yeasts or animals that has been reported to date. A number of plants, however, seem to have structural homologs of the *NPK1* gene, since *Arabidopsis* is shown to have at least three homologs, *ANP1*, *ANP2* and *ANP3* (Nishihama et al. 1997), and Southern blot analysis shows that nuclear DNAs from other plants have sequences which are hybridized with the *NPK1* sequence (Banno et al. 1993a). Thus, NPK1-related protein kinases may be unique to the plant kingdom.

An alignment of the amino acid sequences of NPK1 with its *Arabidopsis* homologs revealed the existence of five conserved regions (A-E) in the kinase-unrelated regions of these protein kinases (Fig. 1). These regions exhibit several interesting features. Two motifs that may be involved in protein-protein interaction have been found. Regions A and C include an acidic residue-rich motif. Such acidic motifs in transcription factors are shown to be important for transcriptional activation through protein-protein interactions with other factors (Hengartner et al. 1995; Gupta et al. 1996). In addition, computer analysis with the COILS program (Lupas et al. 1991) predicted that region E has a high probability of forming a coiled-coil structure. It has been reported that a number of proteins can form a homodimer or a heterodimer via coiled coils (Lupas 1996). The PEST-like sequence, which is thought to be involved in rapid protein degradation (Rechsteiner and Rogers 1996), resides in region C and its flanking region. It is not conserved in the sequence of ANP3. Regions D and E in NPK1 and ANP1L include a consensus motif for phosphorylation by the Cdc2 kinase (S/T-P-X-K/R; Nigg 1993) but ANP2 and ANP3 has only a single motif in regions E and D, respectively. Recent results in our laboratory showed that activated Cdc2 from a seastar phosphorylated NPK1 protein *in vitro* (unpubl. data). Region E contains a stretch of sequence similar to the nuclear

**Fig. 1.** Structural features of NPK1-related MAPKKKs. **A** Alignment of amino acid sequences of tobacco NPK1 and its *Arabidopsis* homologs ANP1L, ANP1S, ANP2 and ANP3. The most carboxyl-terminal five amino acid residues of ANP1S are shown after the position of the last residue of the sequence that is common in ANP1L and ANP1 (*arrow*). Residues that are conserved in three out of the four proteins (except ANP1S) are shown by *white letters on black backgrounds*. The kinase domains are *double-underlined*. The short conserved stretches (regions *A–E*) in the four proteins are *underlined*. The acidic regions, the PEST-like sequence, the sequence similar to the nuclear localization signal and consensus motifs for phosphorylation by Cdc2 kinase are indicated by *Acidic*, *PEST*, *NLS* and *Cdc2*, respectively, over the NPK1 sequence. The last residue of NPK1ΔC is indicated by an *arrowhead* over the NPK1 sequence. **B** Schematic representation of the structure of the NPK1 protein. A *shaded box* and *black boxes* represent the kinase domain and the conserved regions *A-E*, respectively. Predicted coiled-coil regions *Coil*, as well as several motifs shown in **A**, are indicated *over* or *under* the box. Minus signs (–) represent the acidic regions

localization signal. These distinct features of the conserved regions in the kinase-unrelated domains of NPK1-related protein kinases predict interactions of these proteins with unknown factors. As described below, the kinase-unrelated region of NPK1 is shown to be the site where it interacts with its activator protein. Therefore, we designate this domain as a regulatory domain.

## 4
## Activity of ANP1 May be Regulated by Differential Splicing

From a single *ANP1* gene, two types of transcripts are generated: ANP1L and ANP1S transcripts (Nishihama et al. 1997). The former transcript had an intron-like sequence in the coding region for the regulatory domain, while the latter did not include such an intron-like sequence. The ANP1L transcript encodes a putative protein with both the kinase and regulatory domains, resembling NPK1 (Fig. 1A). The ANP1S transcript, however, encodes only the amino-terminal kinase domain of ANP1L because removal of the intron-like sequence generates an in-frame termination codon immediately after the site of the removal (Fig. 1A). These two types of transcripts are probably generated by differential splicing because of the presence of the intron-like sequence in the coding region. Such an intron-like sequence is not present in the coding region of *NPK1*.

NPK1ΔC has higher activity to complement the *bck1* mutation of yeast than NPK1 (see the previous Sect.). Examination of activities of ANP1L and ANP1S with the ste11 mutant of yeast showed that the ability of ANP1S to activate the yeast STE7 MAPKK is greater than that of ANP1L (Nishihama et al. 1997). Thus, activity of the ANP1 protein kinase is probably controlled in part by differential splicing, although it is not known whether such splicing can be regulated by a certain signal. Recently, MEKK1 which is a mouse member of the STE11-type MAPKKKs was reported to be activated by proteolysis in response to an apoptosis signal (Cardone et al. 1997; Deak et al. 1998; Widmann et al. 1998). Upon proteolysis, a portion of the regulatory domain of MEKK1 is removed to generate a truncated protein. Although mechanisms to remove the negative regulatory regions of ANP1 and MEKK1 are different, activated forms of the MAPKKKs can be commonly produced in both cases. Thus, removal of a regulatory domain from a protein kinase may be one of mechanisms for activation of a group of protein kinases.

## 5
## Expression Pattern of the NPK1 Gene

As described in the previous Sect., Northern blotting analysis suggested that the *NPK1* gene is transcribed in proliferating cells. Sites of the expression has

**Fig. 2.** Histochemical localization of GUS activity of a
seedling of transgenic tobacco that has PNPK1 : : GUS.
A 7-day-old seedling of a transgenic tobacco that had the
*NPK1* promoter-fused GUS reporter gene (PNPK1 : : GUS)
was subjected to staining with X-glucuronide (see
Nakashima et al. 1998 for details). *Blue* regions indicate
sites of the *NPK1* expression. *Bar* 1 mm

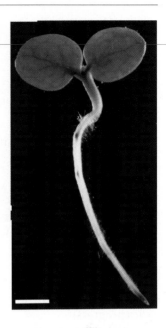

been examined more precisely by *in situ* RNA hybridization or by analyzing
expression patterns of the β-glucuronidase (GUS) reporter gene fused to the
promoter region of the *NPK1* gene in transgenic plants (Nakashima et al.
1998). Embryos from dry seeds exhibit no *NPK1* expression. *NPK1* expression
is first detected in tips of the cotyledons and the radicle during imbibition. In
seedlings, it is detected in the shoot and root apical meristems, surrounding
tissues of the meristems, vascular systems of the hypocotyl and the cotyledons,
and primordia of the lateral roots (Fig. 2). *NPK1* is also expressed in whole
parts of the cotyledons, in which DNA replication and cell division are shown
to occur. As the seedling grows, expression in the cotyledons disappears and
then that in newly emerged first and second leaves appears. Expression of
*NPK1* is detected in all the floral organs in developing flowers; relatively strong
activity is found especially in the stigma, the style, and the ovules in the immature gynoecium. It is also expressed in tips of the adventitious roots which are
formed on the stems of transgenic plants. Thus, expression of *NPK1* is found in
all the meristematic tissues and developing immature organs. Expression of
the *NPK1* gene seems to correlate with the division of cells or with division
competence.

   This correlation has been confirmed by a further experiment, in which a
pattern of distribution of the GUS-positive cells in a transgenic tobacco root
was compared with those in which DNA had replicated (Nakashima et al.
1998). Nuclei of DNA replication-ongoing cells were visualized by incorporation of 5-bromodeoxyuridine (BrdU) followed by staining with anti-BrdU
antibody and FITC-conjugated second antidody. The distribution pattern of

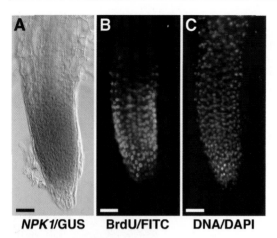

**Fig. 3.** Comparison between distributions of DNA replication-ongoing cells and *NPK1*-expressing cells in a root. A 6.5-day-old seedling of transgenic tobacco that carried PNPK1 : : GUS was incubated in the presence of BrdU for 12 h. The seedling was stained with X-glucuronide to detect *NPK1*-expressing cells (**A**); whole-mount BrdU detection in nuclei of the seedling was carried out with anti-BrdU antibody and FITC-conjugated second antibody (**B**); nuclei were stained with DAPI (**C**). (See Nakashima et al. 1998 for details.) *Bars* 50 μm

**NPK1/GUS      BrdU/FITC      DNA/DAPI**

the GUS-positive cells roughly corresponds to that of the BrdU-incorporating cells (Fig. 3), indicating that the *NPK1* gene is transcribed in cells in which cell cycles progress.

Available data suggest that *NPK1* is expressed probably prior to DNA replication during cell cycle progression (Nakashima et al. 1998). When seeds are allowed to be germinated in the presence of aphidicolin, a DNA polymerase inhibitor, detectable replication of nuclear DNA does not occur in embryos. Under these conditions, *NPK1* is expressed in the apex of the radicle and the cotyledons of embryos. Treatment of seedlings with auxin induces formation of lateral roots, and also induces expression of *NPK1* in the primordia of the lateral roots which are formed in the pericycle. Addition of aphidicolin to a solution containing auxin inhibits the formation of the lateral roots but still induces expression of *NPK1* in the pericycle cells. These results show that expression of *NPK1* can be induced without DNA replication: it probably precedes the entry of the cells into S phase of the cell cycle.

In tobacco cultured cell BY-2, *NPK1* is preferentially transcribed in cells in the logarithmic phase but not in the stationary phase (Banno et al. 1993a). This expression pattern suggests cell cycle-dependent transcription of *NPK1*. Our unpublished data demonstrated such a pattern of transcription during the cell cycle. Northern blot analysis of RNA molecules that were prepared from synchronized BY-2 cells showed that the *NPK1* transcript is accumulated from S to M phases and that the amount of the transcript slightly increases at the M phase and rapidly decreases after the M phase. Western blot analysis revealed almost the same pattern of accumulation of the NPK1 protein as that of the transcript (unpubl. data). Thus, although a peak of the abundance of the NPK1 protein is found at M phase, it does not depend tightly on a particular phase of the cell cycle. A similar pattern of expression is also observed in the cases of plant CDK genes (Hemerly et al. 1993; Setiady et al. 1996).

# 6
## Upstream and Downstream Factors of NPK1

Structural features of a protein often hint at its physiological functions, although they must be experimentally substantiated. Having such an idea, we attempted to isolate tobacco cDNAs which could encode an activator(s) of NPK1. In order to clone cDNA(s) for the activator(s), we took advantage of the cloning system with budding yeast which had been used to isolate a stimulator of activity of Raf MAPKKK (Irie et al. 1994). Details of the system should be referred to in Machida et al. (1998). Eventually, two cDNAs were obtained. Proteins which are predicted to be encoded by these cDNAs have amino acid sequences homologous to those of the motor domains of the kinesins, microtubule-based motor proteins (Vernos and Karsenti 1996; Walczak and Mitchson 1996), and are designated NPK1-activating kinesin-like proteins (NACK1 and NACK2; our unpubl. data).

Immunochemical analyses using yeast cells showed that activity of the NPK1 protein kinase was activated by interacting with NACK1 or NACK2 (our unpubl. data). By using NACK1 and NPK1 cDNAs in yeast cells, we have recently isolated a tobacco cDNA, which was predicted to encode a new member of the MAPKK family (named **Q** in Fig. 4). The activated form of NPK1 had high autophosphorylating activity and phosphorylated the MAPKK protein **Q**.

Subsequently, we attempted to isolate a cDNA for a MAPK protein which can act downstream of MAPKK **Q**, by screening for interacting molecules, and have obtained a candidate cDNA clone (the predicted protein was named **R** in Fig. 4). Accumulation patterns of transcripts of these downstream factors during

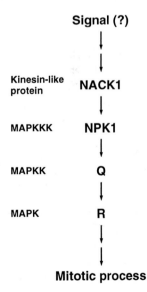

**Fig. 4.** Proposed MAP kinase cascade including NPK1 which is regulated by NACK1 in a tobacco plant. Proteins, **Q** and **R**, represent the newly isolated MAPKK and MAPK, respectively, which are proposed to be placed downstream of NPK1. Presumptive directions of signalling are shown by *arrows*

the cell cycle were similar to those of transcripts of NPK1 (unpublished data). Based on these data, we would like to propose a novel MAP kinase cascade in a plant which can be controlled by the kinesin-like protein NACK1 (Fig. 4).

# 7
# Possible Function of NPK1

Transcripts of the *NACK1* gene were accumulated only at the M phase of the BY-2 cells (our unpubl. data). The promoter region of the *NACK1* gene has M phase-specific *cis* elements, called M-specific activators (MSAs), which are present in promoters of various M phase-specific cyclin genes of various plant species, and confer the M phase-specific transcriptions in plant cells (Ito et al. 1998). As described in the previous Sect., the peak of accumulation of the NPK1 protein is also at the M phase. These observations are in line with an idea that these proteins function in a certain mitotic process.

Since NACK1 seems to be a microtubule-associated motor protein, functions of NPK1 and NACK1 proteins are supposed to be related to physiological events which involve microtubules. To test this assumption, we examined subcellular localization of these proteins. Our preliminary results of immunofluorescence experiments showed that NPK1 and NACK1 were both localized to the central region of the spindle and the phragmoplast. We speculate that the MAP kinase cascade which includes NACK1 and NPK1 could regulate some mitotic event specific to plant cells, such as phragmoplast assembly or cell plate formation.

# References

Ach RA, Durfee T, Miller AB, Taranto P, Hanley-Bowdoin L, Zambryski PC, Grüissem W (1997) RRB1 and RRB2 encode maize retinoblastoma-related proteins that interact with a plant D-type cyclin and geminivirus replication protein. Mol Cell Biol 17:5077–5086

Asada T, Kuriyama R, Shibaoka H (1996) TKRP125, a kinesin-related protein involved in the centrosome-independent organization of the cytokinetic apparatus in tobacco BY-2 cells. J Cell Sci 110:179–189

Banno H, Hirano K, Nakamura T, Irie K, Nomoto S, Matsumoto K, Machida Y (1993a) *NPK1*, a tobacco gene that encodes a protein with a domain homologous to yeast BCK1, STE11, and Byr2 protein kinases. Mol Cell Biol 13:4745–4752

Banno H, Muranaka T, Ito Y, Moribe T, Usami S, Hinata K, Machida Y (1993b) Isolation and characterization of cDNA clones that encode protein kinases of *Nicotiana tabacum*. J Plant Res 3:181–192

Cardone MH, Salvesen GS, Widmann C, Johnson G, Frisch SM (1997) The regulation of anoikis: MEKK-1 activation requires cleavage by caspases. Cell 90:315–323

Deak JC, Cross JV, Lewis M, Qian Y, Parrott LA, Distelhorst CW, Templeton DJ (1998) Fas-induced proteolytic activation and intracellular redistribution of the stress-signaling kinase MEKK1. Proc Natl Acad Sci USA 95:5595–5600

Doerner PW (1994) Cell cycle regulation in plants. Plant Physiol 106:823–827

Giddings TH Jr, Staehelin LA (1991) Microtubule-mediated control of microfibril deposition: a re-examination of the hypothesis. In: Lloyd CW (ed) The cytoskeletal basis of plant growth and form. Academic Press, London, pp 85–99

Grafi G, Burnett RJ, Helentjaris T, Larkins BA, DeCaprio JA, Sellers WR, Kaelin WG Jr (1996) A maize cDNA encoding a member of the retinoblastoma protein family: involvement in endo-reduplication. Proc Natl Acad Sci USA 93:8962–8967

Gu X, Verma DPS (1996) Phragmoplastin, a dynamin-like protein associated with cell plate formation in plants. EMBO J 15:695–704

Gupta R, Emili A, Pan G, Xiao H, Shales M, Greenblatt J, Ingles CJ (1996) Characterization of the interaction between the acidic activation domain of VP16 and the RNA polymerase II initiation factor TFIIB. Nucleic Acids Res 24:2324–2330

Hanks SK, Quinn AM (1991) Protein kinase catalytic domain sequence data-base: Identification of conserved features of primary structure and classification of family members. In: Hunter T, Sefton BM (eds) Methods in enzymology. Protein phosphorylation, vol 200. Academic Press, San Diego, pp 38–62

Hemerly AS, Ferreira P, de Almeida Engler J, Van Montagu M, Engler G, Inzé D (1993) cdc2a expression in Arabidopsis is linked with competence for cell division. Plant Cell 5:1711–1723

Hengartner CJ, Thompson CM, Zhang J, Chao DM, Liao SM, Koleske AJ, Okamura S, Young RA (1995) Association of an activator with an RNA polymerase II holoenzyme. Genes Dev 9:897–910

Irie K, Takase M, Lee KS, Levin DE, Araki H, Matsumoto K, Oshima Y (1993) MKK1 and MKK2, which encode Saccharomyces cerevisiae mitogen-activated protein kinase-kinase homologs, function in the pathway mediated by protein kinase C. Mol Cell Biol 13:3076–3083

Irie K, Gotoh Y, Yashar BM, Errede B, Nishida E, Matsumoto K (1994) Stimulatory effects of yeast and mammalian 14-3-3 proteins on the Raf protein kinase. Science 265:1716–1719

Ito M, Iwase M, Kodama H, Lavisse P, Komamine A, Nishihama R, Machida Y, Watanabe A (1998) Transcription of plant B-type cyclin genes is regulated by the M-phase specific cis-acting activator elements similar to mammalian Myb binding sites. Plant Cell 10:331–341

Kamada Y, Jung US, Piotrowski J, Levin DE (1995) The protein kinase C-activated MAP kinase pathway of Saccharomyces cerevisiae mediates a novel aspect of the heat shock response. Genes Dev 9:1559–1571

Lange-Carter CA, Pleiman CM, Gardner AM, Blumer KJ, Johnson GL (1993) A divergence in the MAP kinase regulatory network defined by MEK kinase and Raf. Science 260:315–319

Lauber MH, Waizenegger I, Steinmann T, Schwarz H, Mayer U, Hwang I, Lukowitz W, Jürgens GJ (1997) The Arabidopsis KNOLLE protein is a cytokinesis-specific syntaxin. J Cell Biol 139:1485–1493

Lee KS, Levin DE (1992) Dominant mutations in a gene encoding a putative protein kinase (BCK1) bypass the requirement for a Saccharomyces cerevisiae protein kinase C homolog. Mol Cell Biol 12:172–182

Lee KS, Irie K, Gotoh Y, Watanabe Y, Araki H, Nishida E, Matsumoto K, Levin DE (1993) A yeast mitogen-activated protein kinase homolog (Mpk1p) mediates signalling by protein kinase C. Mol Cell Biol 13:3067–3075

Levin DE, Fields FO, Kunisawa R, Bishop JM, Thorner J (1990) A candidate protein kinase C gene, PKC1, is required for the S. cerevisiae cell cycle. Cell 62:213–224

Lupas A (1996) Coiled coils: new structures and new functions. Trends Biochem Sci 21:375–382

Lupas A, Dyke MV, Stock J (1991) Predicting coiled coils from protein sequences. Science 252:1162–1164

Machida Y, Nakashima M, Morikiyo K, Banno H, Ishikawa M, Soyano T, Nishihama R (1998) MAPKKK-related protein kinase NPK1: regulation of the M phase of plant cell cycle. J Plant Res 111:243–246

Nakashima M, Hirano K, Nakashima S, Banno H, Nishihama R, Machida Y (1998) The expression pattern of the gene for NPK1 protein kinase related to mitogen-activated protein kinase kinase kinase (MAPKKK) in a tobacco plant: correlation with cell proliferation. Plant Cell Physiol 39:690–700

Nigg EA (1993) Cellular substrates of p34$^{cdc2}$ and its companion cyclin-dependent kinases. Trends Cell Biol 3:296–301

Nishihama R, Banno H, Shibata W, Hirano K, Nakashima M, Usami S, Machida Y (1995) Plant homologues of components of MAPK (mitogen-activated protein kinase) signal pathways in yeast and animal cells. Plant Cell Physiol 36:749–757

Nishihama R, Banno H, Kawahara E, Irie K, Machida Y (1997) Possible involvement of differential splicing in regulation of the activity of *Arabidopsis* ANP1 that is related to mitogen-activated protein kinase kinase kinases (MAPKKKs). Plant J 12:39–48

Nishizuka Y (1992) Intracellular signaling by hydrolysis of phospholipids and activation of protein kinase C. Science 258:607–614

Rechsteiner M, Rogers SW (1996) PEST sequences and regulation by proteolysis. Trends Biochem Sci 21:267–271

Setiady YY, Sekine M, Hariguchi N, Kouchi H, Shinmyo A (1996) Molecular cloning and characterization of a cDNA clone that encodes a Cdc2 homolog from *Nicotiana tabacum*. Plant Cell Physiol 37:369–376

Staehelin LA, Hepler PK (1996) Cytokinesis in higher plants. Cell 84:821–824

Toda T, Shimanuki M, Yanagida M (1993) Two novel protein kinase C-related genes of fission yeast are essential for cell viability and implicated in cell shape control. EMBO J 12:1987–1995

Torres L, Martin H, Garcia-Saez MR, Arroyo J, Molina M, Sanchez M, Nombela C (1991) A protein kinase gene complements the lytic phenotype of *Saccharomyces cerevisiae lyt2* mutants. Mol Microbiol 5:2845–2854

Vernos I, Karsenti E (1996) Motors involved in spindle assembly and chromosome segregation. Curr Opin Cell Biol 8:4–9

Walczak CE, Mitchison TJ (1996) Kinesin-related proteins at mitotic spindle poles: function and regulation. Cell 85:943–946

Widmann C, Gerwins P, Johnson NL, Jarpe MB, Johnson GL (1998) MEK kinase 1, a substrate for DEVD-directed caspases, is involved in genotoxin-induced apoptosis. Mol Cell Biol 18:2416–2429

Williamson RE (1991) Orientation of cortical microtubules in interphase plant cells. Int Rev Cytol 129:135–206

Xie Q, Sanz-Burgos AP, Hannon GJ, Gutierrez C (1996) Plant cells contain a novel member of the retinoblastoma family of growth regulatory proteins. EMBO J 15:4900–4908

# Mitogen-Activated Protein Kinase and Abscisic Acid Signal Transduction

Sjoukje Heimovaara-Dijkstra, Christa Testerink, and Mei Wang[1]

**Summary:** The phytohormone abscisic acid (ABA) is a classical plant hormone, responsible for regulation of abscission, diverse aspects of plant and seed development, stress responses and germination.

It was found that ABA signal transduction in plants can involve the activity of type 2C-phosphatases (PP2C), calcium, potassium, pH and a transient activation of MAP kinase. The ABA signal transduction cascades have been shown to be tissue-specific, the transient activation of MAP kinase has until now only been found in barley aleurone cells.

However, type 2C phosphatases are involved in the induction of most ABA responses, as shown by the PP2C-deficient *abi*-mutants. These phosphatases show high homology with phosphatases that regulate MAP kinase activity in yeast. In addition, the role of farnesyl transferase as a negative regulator of ABA responses also indicates towards involvement of MAP kinase in ABA signal transduction. Farnesyl transferase is known to regulate Ras proteins, Ras proteins in turn are known to regulate MAP kinase activation. Interestingly, Ras-like proteins were detected in barley aleurone cells.

Further establishment of the involvement of MAP kinase in ABA signal transduction and its role therein, still awaits more study.

# 1
## Introduction: Facts and Speculation

Mitogen activated protein kinase (MAP kinase) is a universal signalling tool in all eukaryotes. In plants, MAP kinase has been proposed as a mediator of various signalling events triggered by wounding, abiotic stresses, plant hormones and pathogens (e.g. Mizoguchi et al. 1996; Ligterink et al. 1997; Hirt 1997; Somssich 1997; Wilson et al. 1997; Zhang et al. 1998). One of the phytohormones, abscisic acid (ABA), induces a short, transient activation of MAP

---

[1] Center for Phytotechnology RUL/TNO, TNO Department of Plant Biotechnology, Wassenaarseweg 64, 2333 AL Leiden, The Netherlands

Results and Problems in Cell Differentiation, Vol. 27
Hirt (Ed.): MAP Kinases in Plant Signal Transduction
© Springer-Verlag Berlin Heidelberg 2000

kinase in barley aleurone protoplasts (Knetsch et al. 1996). So far, this is the only publication on the role of MAP kinase in ABA action, but there are several reasons to speculate on MAP kinase involvement in other ABA signalling routes. First of all, there is a link between the physiological functions of ABA in plants and the physiological processes in which MAP kinases act in animals and yeast. Secondly, there is emerging evidence that ABA signal transduction routes share links with the currently known MAP kinase cascades, like the involvement of type 2C protein phosphatases (e.g. Shiozaki and Russel 1995).

It is becoming increasingly clear that the signal transduction cascades of ABA are as diverse as its action. Different tissues respond differently to ABA. Responses to environmental stresses are sometimes achieved via ABA, but are sometimes independent. Here we evaluate the possible role of MAP kinases in the complex picture of ABA action.

# 2
# Physiology of ABA

## 2.1
## Physiological Function of ABA

The phytohormone abscisic acid (ABA) is one of the five classical plant hormones. It was first discovered in the 1960s as the hormone responsible for regulation of abscission. Since then it has been discovered that ABA has a number of important roles in plant growth and development, such as an endogenous anti-transpirant, a mediator of stress responses, an inducer of cell differentiation and a regulator of seed development and germination.

In seed development, ABA is responsible for the accumulation of storage proteins and for survival of the seed tissues during the dehydration of the ripening seed (Balboa-Zavala and Dennis 1977; Black 1983). During development and after completion of seed ripening, ABA plays a role in the regulation of germination; its presence can prevent germination and can induce dormancy of the seed (e.g. Black 1983). In the course of germination itself, it is able to counteract the induction of hydrolytic enzymes that are induced by gibberellin (GA).

Another role of ABA in seed development seems to be the induction of differentiation, especially with regard to embryonic development; it was found that the presence of ABA is essential for the differentiation of calli of *Penniseteum* or cell-culture of carrot into embryonic growth ( Rajasekaran et al. 1987; Kiyouse et al. 1992). ABA also induces the differentiation of photo-synthetic tissue in sedge, required for the C-4 pathway (Ueno 1998)

The role of ABA during seed ripening has close overlap with the role of ABA in stress response: protection against dehydration. Apart from its role in the survival of this "internal" stress, ABA is also involved in plant responses to external stresses such as drought or water stress (e.g. Meurs et al. 1992), cold stress (e.g. Galiba et al. 1993) and salt stress (e.g. Singh et al. 1987). In addition, ABA regulates the closure of stomata to prevent excess water loss and it induces several genes involved in protection against dehydration damage, such as aldose reductase (Bartels et al. 1991). Most of these abiotic stresses can essentially be classified as dehydration stress or osmotic stress.

## 2.2
## Analogies Between ABA Function and MAP Kinase Action in Plant, Yeast and Vertebrate Physiology

Both the regulation of osmotic response/hypertonic shock and cell cycle control/differentiation are physiological processes in which MAP kinase signalling is known to play a pivotal role in other eukaryotes (Herskowitz 1995, Levin and Errede 1995; Marshal, 1995; Bokemeyer et al. 1996; Mizoguchi et al. 1996).

Another link between ABA-action and known MAP kinase-function in eukaryotes is the regulation of activation of quiescent cells. MAP kinases act in several signalling mechanisms to break cell-cycle arrest (Ruderman 1993). ABA acts as an inhibitor of such activation during onset of germination of seeds. GA, the counterpart of ABA in regulation of germination, activates the quiescent cells of the embryo during germination and is found to induce the expression of two MAP kinase homologues in *Avena* (Huttley and Philips 1995). An effect of ABA on the expression of these genes has not yet been reported.

ABA plays an important role in osmotic responses, cell-cycle control and possibly in differentiation. With regard to these roles and the conservation of the signalling between different eukaryotes, one is tempted to speculate on the role of MAP kinases in ABA signalling in plants. As illustrated in other chapters in this book, several components of the MAP kinase signalling cascade have been identified in plants, and many genes have been cloned, awaiting the discovery of their physiological functions.

## 3
## Signal Transduction Cascades of ABA

### 3.1
### Tissue Specificity of ABA Action

The nature of the cellular responses induced by ABA are dependent on the cell type. Two completely different target cells of ABA are represented by stomatal guard cells and cells of the aleurone layer of cereal grains. Guard cells are specialised to undergo rapid volume changes in response to a number of stimuli, including ABA, in order to regulate gas exchange at the leaf surface. Aleurone cells, on the other hand, are components of a secretory tissue geared to synthesise and secrete hydrolytic enzymes during seed germination. In accordance with there function, ABA plays a role in the osmotic-controlled swelling of the guard cell whereas in aleurone its role is to inhibit both hydrolyse production and apoptosis (Wang et al. 1996). In both cases, the responses elicited by ABA differ, as do the putative signal transduction responses. The signal-transduction components in the two cell-types are similar in that both involve alteration in intracellular free-calcium concentrations (Gilroy et al. 1991; Wang et al. 1991; Gilroy and Jones 1992; Irving et al. 1992; McAinsh et al. 1992), intracellular pH (Gehring et al. 1990; MacRobbie 1991; van der Veen et al. 1992), potassium channel activity (Van Duijn et al. 1997; Blatt and Armstrong 1993), L-malate concentration (Hedrich et al. 1994; Heimovaara-Dijkstra et al. 1995) and the expression of specific genes (e.g. Neill et al. 1993). However, the effect of ABA on these mediators in the different tissues is often contradictory as is illustrated for guard cells versus aleurone cells in table 1. In addition, even in one cell type, signal transduction can go via different routes; Allan et al. (1994) found evidence for two different ABA signal transduction routes in guard cells of *Commelina*, one involving changes in intracellular calcium, the other independent of calcium, depending on the temperature at which the plants were grown.

The tissue specificity of ABA signalling is also illustrated by the tissue-specific expression of the *Era1* gene. The *Arabidopsis era1* mutant, with enhanced responsiveness to ABA during germination, appeared to contain a mutation in farnesyl transferase. mRNA from farnesyl transferase in wild type plants was only detected in the flower buds, suggesting involvement in early seed development (Cutler et al. 1996). Farnesyl transferase is involved in the farnesylation of specific proteins, such as Ras and the γ subunit from heterotrimeric GTP-binding proteins. Farnesylation anchors these proteins to membrane lipids or proteins, thus playing a role in the signal transduction process. Since the *era1* mutant does not have a functional farnesyl transferase and is over sensitive to ABA, this suggests a negetative role of farnesyl transferase in ABA signal transduction.

The signal transduction of stress responses adds yet another dimension to the complex picture; not only is tissue-specificity involved but also the partic-

**Table 1.** Putative messengers of ABA signal transduction in guard cells and aleurone cells, references are given in the text

|                              | Guard cell          | Aleurone             |
| ---------------------------- | ------------------- | -------------------- |
| Cytosolic $[Ca^{2+}]_i$      | Increase/no effect  | Decrease             |
| Intracellular $pH_i$         | Increase            | Increase             |
| K channel activity           | Decrease            | Increase             |
| [L-malate]                   | Increase            | Increase             |
| MAP kinase                   | ?                   | Transient activation |

ular stressor in that tissue can be mediated ABA-dependently or -independently. Shinozaki and Yamaguchi-Shinozaki (1997) propose that at least four independent pathways exist for the signalling of drought and salt stresses, two dependent on ABA and two independent. One of the ABA-independent paths overlaps with the signalling route for cold stress.

## 3.2
## Phosphorylation/Dephosphorylation Events

It is becoming increasingly evident that phosphorylation/dephosphorylation events play a role in the signal transduction cascades of ABA; ABA has been shown to induce MAP kinase in aleurone protoplasts (Knetsch et al. 1996) and the characterisation of the *Arabidopsis abi*1 and 2 mutants has rendered interesting new insights in the importance of phosphatases in ABA signal transduction routes (e.g. Leung et al. 1997). Furthermore, several reports show the effects of kinase and phosphatase inhibitors on ABA signalling (e.g. Heimovaara-Dijkstra et al. 1996; Esser et al. 1997).

## 4
## MAP Kinase Activation by ABA

The strongest evidence that MAP kinase is involved in ABA signal transduction so far published, is the activation of a MAP kinase by ABA in aleurone of barley grain (Knetsch et al. 1996). Knetsch et al. show that ABA can induce a transient activation of a kinase in barley aleurone cells that uses myelin basic protein (MBP) as a substrate. MBP is a known substrate of MAP kinases.

Several pieces of evidence strongly suggest that the MBP kinase in question is a MAP kinase; the ABA induced activity can be immuno-precipitated with an anti-ERK1 antibody (Fig. 1) and the 45 kDa ABA-induced activity on in-gel activity assay (Fig. 2) has approximately the same molecular weight as a band recognised by the ERK1 antibody. Finally, the ABA induced activity can be immuno-precipitated with an anti-phosphotyrosine antibody (Fig. 3)

Recently, new insight in the regulation of MAP kinases has been presented by the study on the interaction between MAP kinase and small GTP binding proteins (York et al. 1998). In humans, the initial activation of MAP kinases requires Ras (Hughes et al. 1997; Vossler et al. 1997). Prolonged activation is achieved by binding of another small GTP-binding protein: Rap1. Interestingly, epidermal growth factor (EGF) and nerve growth factor (NGF) induce different responses via one and the same MAP kinase, but with different activation kinetics; whereas EGF can induce a short transient activation, NGF induced sustained activation of this kinase.

Following this finding, the short transient activation in barley aleurone protoplasts would involve Ras protein. The existence of Ras-like proteins in barley

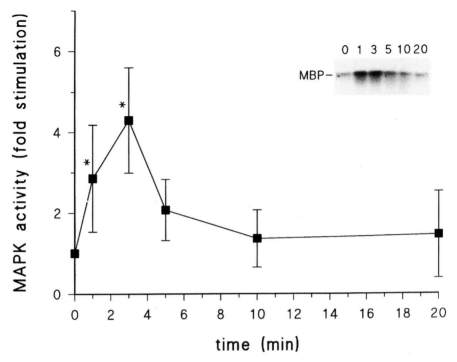

**Fig. 1.** Time course of ABA-induced MAP kinase activity. Barley aleurone protoplasts were treated with ABA. After times indicated, MBP kinase activity was measured in MAP kinase immunoprecipitates. MAP kinase activity is shown as fold stimulation of the MAP kinase activity at time zero. The inset shows MBP phosphorylation of a typical MAP kinase experiment. Mean values of seven independent experiments are shown, * indicate significant difference with t=0, as determined with Student's t-test (p<0.01). Taken from Knetsch et al., 1996

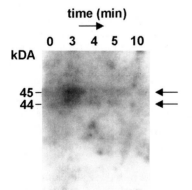

**Fig. 2.** *In gel* MBP kinase activity in barley aleurone protoplasts. Barley aleurone protoplasts were treated with ABA. After times indicated, MBP kinase activity was measured in an in-gel assay as described by Maeda et al., (1996)

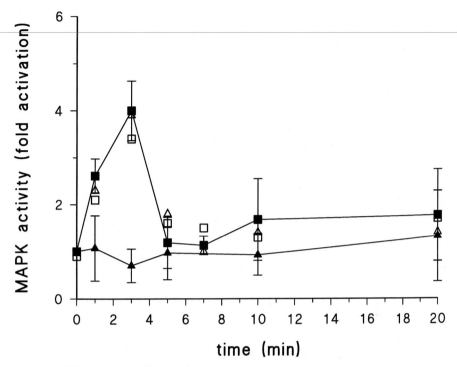

**Fig. 3.** Time course of ABA-induced, anti-phosphotyrosine antibody-precipitated MAP kinase activity. Protoplasts of barley aleurone were treated with ABA for the times indicated. MBP kinase activity was measured in anti-phosphotyrosine antibody immuno-prcipitates (squares) or in the presence of 0.1 mM phosphotyrosine residues (solid triangles) or in the presence of 0.1 mM phosphothreonine residues (open triangles). MAP kinase is presented as fold stimulation of the MAP kinase activity found at time zero. Means of three independent experiments are shown. Taken from Knetsch et al., 1996

aleurone was demonstrated by Wang et al. (1993). The involvement of farnesyl transferase as a negative regulator in ABA signal transduction supports the involvement of GTP binding proteins in ABA signalling, but would contradict with the activation of MAP kinase via Ras.

It will be interesting to see if ABA acts on the activity of MAP kinase in other tissues as well. Knowing the divergence in signal transduction, it is very possible that MAP kinase is not affected in a similar way in other organs/tissues, or is not involved at all. However, if the response would be comparable in other organs/tissues, such a short, transient activation might be difficult to detect. The barley aleurone is a unique tissue as it consists of one cell type and is highly responsive to ABA and GA (Jacobsen and Beach 1985). The fact that this homogeneous population of cells was stimulated as protoplasts, making the signal "available" for each cell at approximately the same time, makes this experimental system unique for the disclosure of such short, transient responses. In intact aleurane layers, this transient response character may not be detected by the same method since aleurone activation is spatially and temporally regulated (e.g. Wang et al. 1998). Even if the same response does exist *in vivo* in guard cells or embryos, it will be very hard to establish this otherwise than at the single cell level.

An indication that other MBP kinases are involved in ABA signal transduction, is presented by Mori and Muto (1997). They found three MBP kinases to be active in guard cell protoplasts from *Vicia faba*, 46, 48 and 49 kDa proteins. The 48 kDa protein is strongly induced by ABA, with a maximum activation reached at 10 min after stimulation. This activity could not be immuno-precipitated with anti-phosphotyrosine antibodies, indicating that it is probably not a MAP kinase. The 46 and 49 kDa MBP kinases could be immuno-precipitated with anti-phosphotyrosine when active, indicating that these might be MAP kinases. Their activity was slightly induced by ABA but seems not to be crucial for ABA-mediated stomatal closure.

Three abiotic stresses that are sometimes mediated by ABA, cold, drought and touch, induce gene expression of the MAP kinase homologue ATMPK3 in *Arabidopsis* within 5 min. (Mizoguchi et al. 1996). It has not been reported whether ABA affects the level of gene expression.

It was found that cold and drought induce activity of MMK4 MAP kinase in alfalfa (Jonak et al. 1996), ATMPK3 is closely related to MMK4 of alfalfa. Salt, heat and ABA did not have an effect on MMK4 activity. Possibly, this MAP kinase acts in the ABA-independent signal-transduction route type 4, as proposed by Shinozaki and Yamaguchi-Shinozaki (1997, see sect. 2.1).

# 5
# The Importance of Protein Phosphatases for ABA Signalling

## 5.1
## Disruption of PP2C Genes Leads to ABA Insensitivity

A major breakthrough in ABA signal transduction research was made with the discovery that both the *ABI1* gene (Leung et al. 1994; Meyer et al. 1994) and the *ABI2* gene (Leung et al. 1997) in *Arabidopsis* encode for phosphatases type 2C (PP2C). The *Arabidopsis abi1* and *abi 2* mutants are impaired in ABA responses in several tissues, *abi2* has a very similar phenotype to *abi1* with the exception of its responses to drought rhizogenesis and the induction of certain genes. (Leung et al. 1997). Both mutants show decreased phosphatase 2C activity (Bertauche et al. 1996).

## 5.2
## ABI1 and ABI2 Phosphatases are Homologous
## to a Regulator of MAP Kinase Pathways in Yeast and Alfalfa

The *ABI1* and *2* (PP2C) genes of *Arabidopsis* show high homology (approx. 42% identity in 200 residues of their central domain) with the MP2C gene from alfalfa (Meskiene et al. 1998). The similarity between the different PP2C's is restricted to the central, catalytic domain, indicating that their substrate is conserved, but the regulation of the phosphatase diverges.

The MP2C gene has been shown to be able to inactivate a yeast MAPKKK (STE 11) responsible for osmo-sensing and there is strong evidence that MP2C can inhibit the stress-induced SAMK pathway (Meskiene et al. 1998). This finding is in accordance with the common function of type 2C phosphatases in fission and budding yeast: negative regulation of MAPK pathways responsible for adaptation to abiotic stresses (Shiozaki and Russel 1995).

## 5.3
## Where Do the Type 2C Phopshatases Fit In?

The fact that the *ABI1* and *2* mutants affect most ABA responses suggests that they act early in ABA signal transduction routes at a point before the divergence into tissue specific routes (e.g. Koornneef 1998). Alternatively, they could be a component which is shared in different routes, possibly with different receptors (e.g. Allan and Trewavas 1994). Interestingly, both over-expression of the *ABI1* and *ABI2* PP2C proteins and inhibition of PP2C activity blocks ABA-dependent gene-expression in *Arabidopsis* leaf protoplasts (Sheen 1998). Sheen suggests that *ABI1*, *ABI2* and AtPP2C, a third PP2C from *Arabidopsis*, have redundant functions as negative regulators of ABA signal transduction.

The *ABI1*-encoded phosphatase contains an EF-hand at it's N terminal domain, suggesting a regulatory role for calcium. However, no calcium binding, nor sensitivity to calcium was detected (Sheen 1998). *ABI1* activity is very sensitive for proton and magnesium concentration (Leube et al. 1998). This sensitivity for pH might present a clue for the interaction of the ABA-induced pH shifts and the role of *ABI1* in ABA signal transduction. The increase of the intracellular pH from 7.08 to 7.21 in aleurone (Van der Veen et al. 1992) and with 0.05–0.1 pH unit in guard cells (Gehring et al. 1990) would mean a significant increase of the phosphatase 2C activity at 0.4 M $Mg^{2+}$ (average value for cytosol Leube et al. 1998). If ABA, directly or indirectly, could activate the PP2C enzyme, this enzyme would be expected to inactivate MAP kinase, possibly as a negative feedback mechanism. Alternatively, the activation of PP2C could also play a role in the negative regulation by ABA of other signal transduction routes like those for GA.

Studies using kinase and phosphatase inhibitors give additional evidence for the involvement of phosphatases in ABA signalling. In broad bean and *Pisum sativum* guard cells, okadaic acid, a phosphatase type 1/2A inhibitor, enhances both slow anion channel activity and ABA-induced guard cell closure. K252a, a kinase inhibitor, inhibits both responses (Schmidt et al. 1995, Esser et al. 1997, Hey et al. 1997). However, Pei et al. (1997) found contradictory results in guard cells of *Arabidopsis*; okadaic acid inhibited ABA-activation of anion channels. Whatever the mechanism underlying these results is, they suggest the involvement of several types of phosphatases and at least one kinase. It has to be mentioned that no thorough study on the specificity of the inhibitors used in plants is yet available. The same holds true for another inhibitor, phenyl arside oxide, which in animals specifically inhibits tyrosine phosphatases (Garcia-Moralez et al. 1990).

Heimovaara-Dijkstra et al. (1996) found that inhibition of tyrosine phosphatases with phenylarsine oxide in barley aleurone protoplasts indeed caused hyperphosphorylation of several proteins. Treatment of these protoplasts with PAO eliminated ABA induction of gene expression and ABA-induced MAP kinase activation (Knetsch et al. 1996). Tyrosine phosphatase activity is known to be essential for MAP kinase activation by EGF (Zhao et al. 1995). Recently, an *Arabidopsis* tyrosine phosphatase was cloned whose expression is induced by salt stress but eliminated by cold treatment (Xu et al. 1998). A *Clamydomonas* gene homologous to the human stress-induced tyrosine phosphatase has been found to be able to inactivate MAP kinases from tobacco and alfalfa (MMK2 and MSK7; Haring et al. 1995). Further research will clarify whether these enzymes and activities play a role in the regulatory activities of ABA.

# 6
# Conclusions

ABA has been shown to induce MAP kinase activity in barley aleurone. The activation seems to be relevant for the induction of specific ABA-induced gene-expression (rab gene: responsive to ABA). Both the transient MAP kinase activation and the induction of rab-gene expression by ABA can be inhibited by PAO, a tyrosine phosphatase inhibitor. This suggests that an animal-like tyrosine phosphatase is acting upstream of this specific MAP kinase (Knetsch et al. 1996). In analogy with animal MAP kinase regulation, Ras might be involved in the activation of MAP kinase (e.g. Hughes et al. 1997).

The activation of MAP kinase by ABA in aleurone is the only evidence so far for involvement of MAP kinase in ABA signal transduction. However, as stated in this chapter, there are results which suggest the involvement of MAP kinase cascades in other ABA responses. The type 2C phosphatases, which have significant homology with the phosphatases that regulate MAP kinase in yeast, and are essential for most ABA responses, provide an especially interesting link.

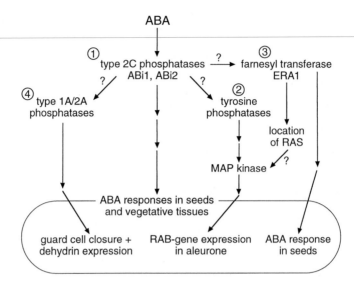

**Fig. 4.** Tentative scheme of possible components of MAP kinase cacades in ABA signalling: Numbers indicate following references1: Leung et al., 1997; 2: Knetsch et al., 1996; 3: Cutler et al., 1996; 4: e.g. Hey et al., 1997

Finally, farnesyl transferase is supposed to regulate Ras protein function which in its turn regulates MAP kinase activation. These responses are presented in a tentative scheme in which PP2C's are placed before the branching point (Fig. 4).

All this leads us to think that there is reason to speculate on MAP kinase involvement in ABA action other than in barley aleurone alone. However, the complexity of ABA signal transduction suggests that if involved at all, it's place and role in the different cascades will be as diverse as ABA action itself.

*Acknowledgements.* We would like to thank Bert van Duijn for critically reading the manuscript. We express our gratitude to Peter Hock for the lay-out of some figures and to Marco Vennik for technical assistance.

# References

Allan AC, Fricker MD, Ward JL, Beale MH, Trewavas AJ (1994) Two transduction pathways mediate rapid effects of abscisic acid in *commelina* guard cells. Plant Cell 6:1319–1328

Allan AC, Trewavas AJ (1994) Abscisic acid and gibberellin perception: inside or out? Plant Physiol 104:1107–1108

Balboa-Zavala O, Dennis FG (1977) ABA and the onset of dormancy in apple. J Am Soc Hortic Sci 102:633–637

Bartels D, Engelhardt K, Roncarati R, Schneider K, Rotter M, Salamini F. (1991) An ABA and GA modulated gene expressed in the barley embryo encodes an aldose reductase related protein. EMBO J 10:1037–1043

Bertauche N, Leung J, Giraudat J (1996) protein phosphatase cativity of abscisic acid insensitive 1 (ABI1) protein from *Arabidopsis* thaliana. Eur J Biochem 341:193–200

Black M (1983) Abscisic acid in seed germination and dormancy. In: Addicott FT (ed) Abscisic Acid, Preager Publishers New York, pp 337–363

Blatt MR, Armstrong F (1993) $K^+$ channels of stomatal guard cells: abscisic-acid evoked control of the outward-rectifier mediated by cytoplasmic pH. Planta 191:330–341

Bokemeyer D, Sorokin A, Du M (1996) Mutiple intracellular MAP kinase signaling cascades. Kidney Int 49:1187–1198

Cutler S, Ghassemian M, Bonetta D, Cooney S, McCourt P (1996) A protein farnesyl transferase involved in abscisic acid signal transduction in *Arabidopsis*. Science 273:1239–1241

Esser JE, Liao YJ, Schroeder JI (1997) Characterization of ion channel modulator effects on ABA- and malate-induced stomatal movements: strong regulation by kinase and phosphatase inhibitors, and relative insensitivity to mastoparans. J Exp Bot 48:539–550

Galiba G, Tuberosa R, Kocsy G, Sutka J (1993) Involvement of chromosomes 5A and 5D in cold-induced abscisic acid accumulation and in frost tolerance of wheat calli. Plant breeding 110: 237–242

Garcia-Moralez P, Minami Y, Luong EL, Klausner RD, Samelson LE (1990) Tyrosine phosphorylation in T-cells is regulated by phosphatase activity: studies with phenylarsine oxide. Proc Natl Acad Sci USA 87:9255–9259

Gehring CA, Irving HR, Parish RW (1990) Effects of auxin and abscisic acid on cytosolic calcium and pH in plant cells. Proc Natl Acad Sci USA 87:9645–9649

Gilroy S, Fricker MD, Read ND, Trewavas AJ (1991) Role of calcium in signal transduction of *Commelina* guard cells. Plant Cell 3:333–344.

Gilroy S, Jones RL (1992) Gibberellic acid and abscisic acid coordinately regulate cytoplasmic calcium and secretory activity in barley aleurone protoplasts. Proc Natl Acad Sci USA 89: 3591–3595

Haring MA, Siderius, M, Jonal, C, Hirt H, Walton, K.M., Musgrave A (1995) Tyrosine phosphatase signalling in a lower plant: cell cycle and oxidative stress-regulated expression of the *Clamydomonas eugametos* VH-PTP13 gene. Plant J 7:981–988

Hedrich R, Marten I, Lohse G, Dietrich P, Winter H, Lohaus G, Heidt H-W (1994) Malate-sensitive anion channels enable guard cells to sense changes in the ambient $CO_2$ concentration. Plant J 6:741–748

Heimovaara-Dijkstra S, Mundy J, Wang M (1995) The effect of intracellular pH on the regulation of the Rab 16A and the -amylase 1/6-4 promoter by abscisic acid and gibberellic acid. Plant Mol Biol 27:815–820

Heimovaara-Dijkstra S, Nieland TJF, Van der Meulen RM, Wang M (1996) Abscisic acid-induced gene-expression requires the activity of protein(s) sensitive to the protein-tyrosine phosphatase inhibitor phenylarsine oxide. Plant Grow Regu 18:115–123

Herskowitz I (1995) Map kinase pathways in yeast: for mating and more. Cell 80:187–197

Hey JH, Bacon A, Burnett E, Neill SJ (1997) Abscisic acid signal transduction in epidermal cells of *Pisum sativum* L argenteum: both dehydrin mRNA accumulation and stomatal responses require protein phosphorylation and dephosphorylation. Planta 202:85–92

Hirt H (1997) Multiple roles of MAP kinases in plant signal transduction. Trends Plant Sci 2:11–15

Hughes PE, Renshaw MW, Pfaff M, Forsyth J, Keivens VM, Schwarts MA, Ginsberg MH (1997) Suppression of integrin activation: a novel function of a ras/raf-initiated MAP kinase pathway. Cell 88:521–530

Huttly AK, Phillips A (1995) Gibberellin-regulated expression in oat aleurone cells of two kinases that show homology to MAP kinase and a ribosomal protein kinase. Plant Mol Biol 27: 1043–1052

Irving HR, Gehring CA, Parish RW (1992) Changes in cytosolic pH and calcium of guard cells precede stomatal movements. Proc Natl Acad Sci USA 89:1790–1794

Jacobsen JV, Beach LR (1985) Control of transcription of -amylase and r-RNA genes in barley aleurone protoplasts by gibberellin and abscisic acid. Nature 316:275–277

Jonak C, Kiegerl S, Ligterink W, Barker PJ, Huskisson BS, Hirt H (1996) Stress signalling in plant: A mitogen-activated protein kinase pathway is activated by cold and drought. Proc Natl Acad Sci USA 93:11274–11279

Kiyosue T, Nakajima M, Yamaguchi I, Satoh S, Kamada H, Harada H (1992) Endogenous levels of abscisic acid in embryogenic cells, non-embryogenic cells and somatic embryos of carrot (*Daucus carota* L.). Biochem Physiol Pflanzen 188:343–347

Knetsch ML, Wang M, Snaar-Jagalska BE, Heimovaara-Dijkstra S (1996) Abscisic acid induced mitogen-activated protein kinase activation in barley aleurone protoplasts. Plant Cell 8:1061–1067

Koornneef M, Leon-Kloosterziel KM, Schwartz SH, Zeevaart JAD (1998) The genetic and molecular dissection of abscisic acid biosynthesis and signal transduction in *arabidopsis*. Plant Physiol Biochem 36:83–89

Leung J, Bouvier-Durand M, Morris PC, Guerrier D, Chefdor F, Giraudat J (1994) Arabidopsis ABA response gene ABI1: features of a calcium modulated protein phosphatase. Science 264 (5164):1448–1452

Leung J, Merlot, S, Giraudat J (1997) The *arabidopsis* abscisic acid insensitive (*abi2*) and *(abi1)* genes encode homologous protein phosphatase 2C involved in abscisic acid signal transduction Plant Cell 9:759–771

Leube MP, Grill E, Amrhein N (1998) ABI1 of *Arabidopsis* is a protein serine/threonine phosphatase highly regulated by the proton and magnesium ion concentration. FEBS Lett 424: 100–104

Levin DE, Errede B (1995) The proliferation of MAP kinase singaling pathways in yeast. Cur Opin Cell Biol 7:197–202

Ligterink W, Kroj, T, zur Nieden U, Hirt H, Scheel D (1997) Receptor-mediated activation of a MAP kinase in pathogen defense of plants. Science 276:2054–2057

MacRobbie MR (1991) Effect of ABA on ion transport and stomatal regulation. In Abscisic acid: Physiology and biochemistry, pp153–168, WJ Davies, Jones HG eds. Bios Scientific Publishers, Oxford, UK

Maeda, T, Wurgler-Murphy, S.M., Saito, H (1994) A two-component system that regulates an osmosensing MAP kinase cascade in yeast. Nature 369:242–245

Marshall CJ, (1995) Specifity of receptor tyrosine kinase signaling: transient versus sustained extracellular signal-regulated kinase activation. Cell 80:179–185

McAinsh MR, Brownlee C, Hetherington AM (1992) Visualising changes in cytosolic-free-$Ca^{2+}$during the response of stomatal guard cells to abscisic acid. Plant Cell 4:1113–1122

Meada M, Aubry L, Insall R, Gaskins C, Devreotes PN, Firtel RA (1996) Seven helix chemoattractant receptors transiently stimulate mitogen-activated protein kinase in *Dictyostelium*-Role of heterotrimeric G proteins. J Biol Chem 271 (7):3351–3354

Meskiene I, Bögre, L, Glaser W, Balog J, Brandtstötter M, Zwerger K, Ammerer, G, Hirt H (1998) MP2C, a plant protein phosphatase 2C, functions as a negative regulator of mitogen-activated protein kinase pathways in yeast and plants. Proc Natl Acad Sci USA (95) 1938–1943

Meurs, C, Basra, A.S., Karssen CS, van Loon LC (1992) Role of abscisic acid in the induction of dessication tolerance in developing seeds of *arabidopsis* thaliana. Plant Physiol 98:1484–1493

Meyer K, Leube MP, Gill E (1994) A protein phosphatase 2C involved in ABA signal transduction in *Arabidopsis thaliana*. Science 264:1452–1455

Mizoguchi T, Irie K, Hirayama T, Hayashida N, Yamaguchi-Shinozaki K, Matsumoto K, Shiozaki K (1996) A gene encoding a mitogen-activated protein kinase kinase kinase is induced simulatiously with genes for a mitogen-activated protein kinase and a S6 ribosomal protein kinase by touch, cold and water stress in *Arabidopsis thaliana*. Proc Natl Acad Sci USA 93: 765–769

Mori IC, Muto S (1997) Abscisic acid activates a 48-kilodalton protein kinase in guard cell protoplasts. Plant Physiol 113:833–839

Neill SJ, Hey SJ, Barratt DHP (1993) Analysis of guard cell gene expression in *Pisum sativum*. J Exp Bot 44:suppl 3

Pei ZM, Kuchitsu K, ward JM, Schwarz M, Schroeder JI (1997) Differential abscisic acid regulation of cell slow anion channels in *Arabidopsis* wild-type and abi1 abi2 mutants. Plant Cell 9(3):409–423

Rajasekara K, Hein MB, Vasil IK (1987) Endogenous abscisic acid and indo-3-acetic acid and somatic embryogenesis in cultured leaf explants of *Penniseteum purourum* Schum. Plant Physiol 84:47–51

Ruderman JV (1993) MAP kinase and the activation of quiscent cells. Curr. Opinion Cell Biol 5: 207–213

Schmidt C, Schelle I, Liao YJ, Schroeder JI (1995) Strong regulation of slow anion channels and abscisic acid signalling in guard cells by phosphorylation and dephorsphorylation events. Proc Natl Acad Sci USA 92:9535–9539

Sheen, J (1998) Mutational analysis of protein phosphatase 2C involved in abscisic acid signal transduction in higher plants. Proc Natl Acad Sci USA 95:975–980

Shinozaki K and Yamaguchi-Shinozaki K(1997) Gene expression and signal transduction in water-stress response. Plant Physiol 115:327–334

Shiozaki, K and Russel P (1995) Counteractive roles of protein phosphatase 2C (PP2C) and a MAP kinase kinase homologue in the osmoregulation of yeast. EMBO J 14:492–502

Singh NK, LaRosa PC, Handa AK Hasegawa PM and Bressan RA (1987) Hormonal regulation of protein synthesis associated with salt tolerance in plant cells. Proc Natl Acad Sci USA 84: 4108–4112

Somssich IE (1997) MAP kinases and plant defence. Trends Plant Sci 2:406–407

Ueno O (1998) Induction of Kranz anatomy and C-4-like biochemical characteristics in a submerged amphibious plant by abscisic acid. Plant Cell 10:571–583

Van der Veen R, Heimovaara-Dijkstra S, Wang M (1992) Cytosolic alkalinization mediated by abscisic acid is necessary, but not sufficient, for abscisic acid-induced gene expression in barley aleurone protoplasts. Plant Physiol 100:699–705

van Duijn B, Flikweert MT, Wang M (1997) Effects of ABA on inward rectifying potassium conductance in barley aleurone protoplasts from dormant and non-dormant barley grains. Biophys J 72:A336

Vossler MR, Yao H, York RD, Pan MG, Rim CR, Stork PJS (1997) cAMP activates MAP kinase and ElK-1 through a B-Raf- and Rap1-dependent pathway. Cell 89:73–84

Wang M, Van Duijn B, Schram AW (1991) Abscisic acid induces a cytosolic calcium decrease in barley aleurone protoplasts. FEBS Lett 278:69–74

Wang M, Sedee NJA, Heidekamp F, Snaar-Jagalska BE (1993) Detection of GTP-binding proteins in barley aleurone protoplasts. FEBS Lett 329:245–248

Wang M, Oppedijk BJ, Lu X, Van Duijn B, Schilperoort RA (1996) Apoptosis in barley aleurone during germination and its inhibition by abscisic acid. Plant Mol Biol 32:1125–1134

Wang M, Oppedijk BJ, Caspers MPM, Lamers GEM, Boot MJ, Geerling DNG, Bakhuizen B, Meijer AH, Van Duijn B (1998) Spatial and temporat regulation of DNA fragmentation in the aleurone of germinating barley. J Exp Bot 49 (325):1293–1301

Wilson C, Voronin V, Touraev A, Vicente O, Herberle-Bors E (1997) A developmentally regulated MAP kinase activated by hydration in tabacco pollen. Plant Cell 9:2093–2100

Xu Q, Fu H-H, Gupta R, Luan S (1998) Molecular characterisation of a tyrosine-specific protein phosphatase encoded by a stress-responsive gene in *arabidopsis*. Plant Cell 10:849–857

York RD, Yao H, Dillon T, Ellig CL, Eckert SP, McClenskey EW, Stork PJS (1998) Rap1 mediates sustained MAP kinase activation induced by nerve growth factor. Nature 392:622–626

Zhang S, Du H, Klessig DF (1998) Activation of the tabacco SIP kinase by both a cell wall-derived carbohydrate elicitor and purified proteinaceous elicitins from *Phytopthora* spp. Plant Cell 10:435–449

Zhao Z, Tan Z, Wright JH, Diltz CD, Shen S, Krebs EG, Fisher EH (1995) Altered expression of protein-tyrosine phosphatase 2C in 293 cells affects protein tyrosine phosphorylation and mitogen-activated protein kinase activation. J Biol Chem 270:11765–11769

# Signal Transduction of Ethylene Perception

Sigal Savaldi-Goldstein and Robert Fluhr[1]

**Summary:** Ethylene signal transduction pathway regulates various aspects of plant physiology and development. Studies of mutants defective in the ethylene response, has led to the elaboration of key genes involved in the perception of ethylene. Among them are putative ethylene receptors, Raf-like kinases, nuclear-targeted proteins and transcription factors. The gene products share common motifs found in other signaling-cascade pathways in organisms ranging from bacteria to mammals. Recent biochemical studies provide insight into the function and regulation of the components of the ethylene cascade and make ethylene perception a paradigm for signal transduction in multicellular organisms.

## 1
## Introduction

Ethylene plays a direct role in plant development as well as in coordinating the processes of fertilization, senescence, wounding and pathogen response. Both the production and perception of ethylene are modulated by the environment and developmental state of the plant. Phosphorylation processes play a central role in transduction of the ethylene response as have been established by biochemical evidence, the isolation of kinases and characterization of genes that participate in the pathway. Central to our understanding of this pathway was the elucidation of its genetic components. The genetic approach was evidently facilitated by the robustness, at least under laboratory growth conditions, of ethylene-insensitive plants. The apparent dispensability facilitates genetic screening (Ecker 1995; Kieber 1997; McGrath and Ecker 1998). Examples of such screens are the isolation of ethylene signal transduction mutants that include disruption of ethylene-induced distinctive changes in seedling morphology or the loss of ethylene-induced acceleration of fruit ripening (Goeschl et al. 1966; Taylor et al. 1988; Guzman and Ecker 1990). In the dark, the seedlings

---

[1] Weizmann Institute of Science, 76100 Israel

Results and Problems in Cell Differentiation, Vol. 27
Hirt (Ed.): MAP Kinases in Plant Signal Transduction
© Springer-Verlag Berlin Heidelberg 2000

show inhibition of both the root geotropic response and root and hypocotyl elongation, as well as exaggerated apical curvature. The morphological phenomenon is known as 'triple response' and the compact growth structure afforded to seedlings by this response may be beneficial during their emergence. Two major classes of mutations have been isolated in *Arabidopsis*. One class, showing loss of 'triple response' is insensitive to exogenous ethylene such as *etr1* (Bleecker et al. 1988) or *ein*-type (Roman et al. 1995) another class shows a constitutive triple response in the absence of added ethylene, such as *ctr1*.

Ethylene plays a key role in ripening of climacteric fruit such as tomato, avocado and banana. Its production and perception are crucial for the coordination and completion of the ripening process. The *never ripe (Nr)* mutations for ethylene signal transduction in tomato were fortuitously isolated by observing fruit that remained on the vine. Further analysis of these mutants showed the inability of leaves, petioles and abscission-zone tissue to respond to ethylene. Thus, *Nr* seedlings lack a triple response, their leaves do not show an epinastic response (downward curvature of leaves) to the ethylene precursor ACC (1-aminocyclopropane-1-carboxylic acid) and their flowers do not senesce or wilt (Lanahan et al. 1994).

A combination of biochemical and genetic approaches have made the components of ethylene signal transduction one of the best characterized hormone response pathways. The perception pathway begins with ethylene binding by a specific receptor followed by a phospho-relay through two-component signal transducers, thereby activating downstream protein kinase cascades that lead to gene induction.

## 2
## Detection of Phosphorylation Events in Ethylene Signal Transduction

The use of biochemical inhibitors has established a role for phosphorylation in the regulation of both ethylene biosynthesis and action. Direct involvement of phosphorylation events in ethylene action was illustrated when specific polypeptides were shown to undergo transient phosphorylation within 10 min of exposure to ethylene (Raz and Fluhr 1993). The transient phosphorylation was prevented in the presence of kinase inhibitors H-7 and K-252a. At the same time, ethylene-induced chitinase accumulation and lesion formation promoted by ethylene, was inhibited. Inhibition of ethylene-induced changes in the level of ACC oxidase transcript was similarly inhibited by H-7 (Kwak and Lee 1997). Thus, rapid polypeptide phosphorylation is associated with ethylene application. Moreover, phosphorylation events are necessary for ethylene-activated signal transduction which results in the induction of ethylene-sensitive pathogenesis-related (PR) gene expression and microlesion formation. In a similar manner, protein phosphorylation of cultured plant cells was shown to be rapidly and transiently affected by fungal elicitor (Dietrich et al. 1990), and negatively regulated by K-252a (Grosskopf et al. 1990; Felix et al. 1998).

Okadaic acid, a specific inhibitor of animal and plant PP1 and PP2A, increased the general level of protein phosphorylation in intact leaves (Raz and Fluhr 1993). Two types of protein phosphorylation kinetics were detected. A stable increase induced by okadaic acid, and a transient increase induced by ethylene. The transient increase is consistent with attenuation of an initial receptor-like response to a signal molecule, while the stable increase is expected from a constitutive biochemical type of block. In addition, okadaic acid elicited the rapid accumulation of a PR-specific transcript and PR proteins in intact leaves. The induction of PR gene expression by okadaic acid was abrogated by the presence of kinase inhibitors, suggesting that the transduction events requiring kinase and phosphatase activity are probably acting through the same pathway. Similar results were obtained in activation of abscisic acid-dependent genes by okadaic acid (Wu et al. 1997). Hence, at least one stage of the transduction pathway requires kinase activity, which is modified by a mechanism that maintains phosphorylation equilibrium. Alternatively, or in addition, the induction of PR proteins by ethylene could be transduced via negative regulation of PP1/PP2A phosphatase activity.

## 3
## A Survey of Putative Genetic and Molecular Components of the Ethylene Signal Pathway

### 3.1
### Mutants of Ethylene Response

Genetic screens have delineated at least two major classes of mutations in *Arabidopsis*. One class, showing loss of 'triple response' is insensitive to exogenous ethylene such as *etr1* (Bleecker et al. 1988) or *ein*-type (Roman et al. 1995); another class shows a constitutive triple response in the absence of added ethylene, such as *ctr1*. The fact that mutations in the ethylene pathway are of either gain of function-type or loss of function-type and are probably part of one common signaling pathway, allows a conceptual framework of their order to be built up by establishing epistatic relationships. Hence, if in double-mutant progeny one mutant phenotype masks the phenotype of another, the simplest assumption is that the predominant locus acts as a downstream modulator. Thus, double mutants of *etr1* and *ctr1* do not show the ethylene-insensitive etr1 phenotype but rather the constitutive ethylene phenotype of *ctr1*. This positions *ETR1* upstream of *CTR1*. On the other hand, *ein2/ctr1* double mutants show the ethylene-insensitive *ein2* phenotype, indicating that the *EIN2* locus acts downstream of *CTR1*. Extensive analysis of this sort has led to the delineation of a pathway based on genetic epistatic relationships (Fig 1).

## 3.2
## The ETR1 Protein and Related Homologues

The *ETR1* gene was isolated using map-based cloning technology and shown to encode structural motifs reminiscent of the two-component environmental sensors (Chang et al. 1993). Environmental sensors are protein communication modules in bacteria that contain two structural motifs – transmitters and receivers – that can be arrayed in complex combinations (Fig. 1). In the simplest format, a sensor device monitors an environmental variable that activates a histidine kinase (the transmitter). Five domains that characterize the sensor-histidine kinase, including the acceptor histidine, are conserved in the N-terminal portion of ETR1. The second motif in the communication module, called a response regulator, is generally an independent polypeptide and func-

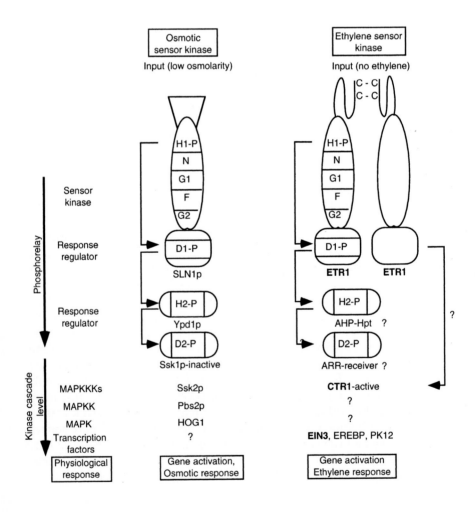

tions to receive the sensor signal. This culminates in phosphoryl transfer from the phosphorylated histidine kinase to an Asp residue in the response regulator. The activated receiver then modulates the activity of a unique C-terminal output domain, commonly a transcriptional regulator. The sensor element can be of the simple type already described or the hybrid type in which the sensor-transmitter kinase domain has an attached regulator domain. Whether simple or hybrid, an additional independent regulator domain exists, hence the descriptive term 'two-component system'.

*ETR1* homologues *ETR2, EIN4, ERS1* and *ERS2* have been cloned in *Arabidopsis*: (Hua et al. 1995; Hua et al. 1998b; Sakai et al. 1998). Similar to *ETR1*, ETR2 and EIN4 are hybrid type kinases. In contrast, *ERS1* and *ERS2* lack the C-terminal receiver domain. Interestingly, variant splicing of *ETR2* transcript was detected which would entail synthesis of an ETR2 gene product without the receiver domain. *ERS*-type homologue was found to encode the *Nr* gene in tomato (Wilkinson et al. 1995). Thus, the simple and hybrid types of ethylene-related histidine kinases coexist in plants.

## 3.3
## Functional Aspects of ETR1 Related Proteins

The N-proximal region is the most highly conserved region shared by all putative ethylene receptor genes discussed above. This region contains invariant

---

**Fig. 1.** Molecular components of the ethylene signal transduction pathway and the yeast osmo-sensing pathway showing phospho-relay and kinase cascade components. The order of the components in *bold* was established by examining epistatic relationships. The positions of putative MAP kinases in the ethylene pathway, and of EREBP and PK12 components remains to be established experimentally. *ETR1* and *SLN1p* are hybrid histidine kinases in which the sensor element and response regulator are combined in one molecule. The model shown depicts active phospho-relay in low osmotic media and in the absence of ethylene for Sln1p and ETR1 respectively. Conserved phosphorylated sites and active-site signatures are indicated in the two-component sensor (*H, N, G, F,* and *D*). The activity of the kinase results in phosphorylation of a conserved His residue (*H1-P*). The phosphate group is subsequently transferred to the response regulator which is phosphorylated on an Asp residue (*D1-P*). In the case of Sln1p, the phosphate is further relayed to His and Asp (*H2-P, D2-P*) residues in *Ypd1p* and *Ssk1p*, respectively. The autophosphorylation of *Sln1p* is activated in low osmolarity followed by phospho-relay that results in a phosphorylated, inactive state of Ssk1p and hence a quiescent state for the MAP kinases related to osmotic regulation (*Ssk2p, Pbs2p, HOG1*). Specific transcription factors related to the osmotic response in yeast have not been identified. The *ETR1* molecule is shown as a homodimer with disulfide linkages in the transmembrane area. *ETR1* is likely to be active in the absence of ethylene as suggested by loss-of -function mutants phenotypes of the ethylene receptors (Hua et al. 1998a). Phospho-relay may occur directly from *ETR1* to *CTR1* or may be coupled through response regulators like the recently described *Arabidopsis* response regulators, *ARR-receiver* and histidine-containing phosphotransfer domain *AHP-Hpt. CTR1* is a negative regulator of elements that are downstream and may repress putative MAP kinase elements that have yet to be identified. *EIN3, EREBP* and *PK12* are nuclear localized and have been implicated in ethylene response and thus may play a role at the level of transcription

cysteine residues that are followed by three putative membrane-spanning domains of 17–23 hydrophobic amino acid stretches each (Fig. 1). This is consistent with a model in which the N-terminus lies on the extracytoplasmic side of the membrane and the C-terminal domain resides on the cytoplasmic side of the membrane. Indeed, specific antibodies detected a dimeric form of ETR1 in membrane fractions from *Arabidopsis* or transgenic yeast expressing ETR1 (Schaller et al. 1995). The dimeric form was found to be sensitive to reductant and dependent on the presence of the proximal extracytoplasmic cysteines that form disulphide bonds. Dimerization of ETR1 fits the paradigm of bacterial two-component signal transduction receptors in which one monomer catalyses the phosphorylation of its neighbor on a histidine residue. For example, both the VirA receptor, responsible for sensing acetosyringone in plant-*Agrobacterium* interactions, and the *E. coli* aspartate receptor, exist as histidine kinase homodimers. Binding of the ligand does not induce the oligomeric structure of these receptors (Milligan and Koshland 1988; Pan et al. 1993). The existence of preformed receptor dimers indicates that the process of dimerization does not play a direct role in signal transduction across the membrane interface, as occurs in some eukaryotic receptor-ligand interactions (Heldin 1995).

The three N-terminal membrane-spanning regions of the putative ethylene receptor gene family, are of particular structural and biological interest. All the mutations identified in these genes, which result in ethylene-insensitivity, reside in the membrane-spanning region. For example, all four *ETR1* mutations in *Arabidopsis* involve single amino acid substitutions that are located in each of the membrane-spanning domains (A31V, I62F, C65Y, A102T). In tomato, a P36L mutation that leads to the ethylene insensitive never-ripe phenotype falls at a comparable position. Ethylene binding to whole plants can be related to ETR1 functionality, because ETR1-C65Y-containing *Arabidopsis* plants show reduced ethylene binding. Transgenic yeast that express wild-type ETR1, but not those that express the ETR1-C65Y mutation bind ethylene (Bleecker et al. 1988; Schaller and Bleecker 1995). These observations suggest that the N-terminal membrane-spanning region of ETR1 fulfills a role as the ethylene-binding site.

The readily reversible binding of ethylene to metals makes it likely that ethylene binding to ETR1 is mediated via coordination to a metal group. Experiments with transgenic yeast expressing ETR1 show that copper associates with membrane extracts and mediates high-affinity ethylene binding activity (Rodriguez et al. 1999). Mutations in either Cys 65 and His 69 of ETR1, abolish the effect suggesting that these residues play a direct role in metal binding. Paradoxically, silver but not other metal ions could partially replace copper. Since silver ions is known to inhibit ethylene perception in plants they can promote ethylene binding yet impair receptor activity. Presumably, a conformational change of state is induced in ETR1 following ethylene-binding and is transduced perpendicularly to the membrane plane. The change in state could have profound effects on the rate of phosphotransfer on the cytoplasmic side of the membrane. Experimental evidence for ligand-dependent movement of trans-membrane components has been obtained for the two component bacterial aspartate sensory receptor. In this case, the receptor was first modified by the substitution of cysteine residues in the transmembrane region. The rate of sulphhydryl cross-linkage was then monitored and shown to change upon addition of aspartate (Lynch and Koshland 1991). A plausible explanation for this rate change is repositioning of the cysteine residues as a result of ligand-activated movement in the transmembrane region.

## 3.4
### Response Regulator Homologues in *Arabidopsis* (ARRs)

Phosphotransfers in bacteria are referred to as two-component systems containing independent sensors and response regulators. While ETR1 is a hybrid type histidine kinase, ERS and Nr are simple types and lack their own response regulators. This leads one to the conclusion that independent response regulators may exist. Furthermore, as will be shown below, the yeast osmosensor Sln1p protein which is a hybrid-type histidine kinase, mediates its response through independent response regulators (Posas et al. 1996). Recently, with the help of the *Arabidopsis* EST database, five cDNA sequences with receiver-like homology to *Synechocystis* response regulators were isolated (Imamura et al. 1998). They are related to the bacterial Che Y response regulator and were shown to be transiently phosphorylated in *E. coli*. The response regulators catalyze the reception of phosphate on its aspartate. The characteristic half-life stability of phosphorylated aspartate component will regulate the efficiency of phospho-relay and ensure its transient state. As an example, the half-life for Che Y is only a few seconds. The involvement of another histidine kinase, (CKI1; Kakimoto 1996) as well as the identification of early induced receiver homologues (IBC6 and IBC7, Brandstatter et al. 1998) in cytokinin signal transduction means that it remains to be established which, if any, of the ARR are involved in ethylene signaling.

## 3.5
### CTR1: A Raf-Like Protein Kinase
### and Candidate Entry to MAP Kinases

*CTR1* has been located, using epistatic relationships, as being downstream from *ETR1* and *EIN4,* and upstream from all other *ein*-type mutations. The gene was isolated with the help of T-DNA-tagged mutants. The C-terminus of CTR1 contains domains typical of protein kinases of the Raf protein kinase family (Kieber et al. 1993). Raf is the retroviral protein that transduces signals from cell-surface receptors to transcription factors which mediate changes in cell growth and differentiation (Gupta and Roberts 1994). Inspection of all mutant alleles of *CTR1* show they would yield truncated or nonfunctional kinases. This indicates that loss of function at this juncture results in constitutive activation of the pathway, suggesting that functional phosphorylating activity of CTR1 represses downstream ethylene response. The N-terminal portion of Raf-type kinases has been hypothesized to regulate the kinase activity, and in the case of CTR1 its N-terminal portion may be the direct or downstream target for ETR1 activity.

In addition to CTR1, which belongs to the MAPKKK family of protein kinases, other members of the MAP kinase family were identified in plants (reviewed in Hirt 1997; Machida et al. 1997). Some of them have been implicated in stress responses like wounding and pathogenesis, known to be mediated by ethylene as well. ERMK (elicitor responsive MAP kinase) phosphorylates MBP (a common substrate for the MAP kinase family) within 5 min (Ligterink et al. 1997). Sequence analysis of the cloned ERMK gene revealed its similarity to other MAP kinases such as the tobacco WIPK (Mizoguchi 1997; Zhang et al. 1998). ERMK is translocated to the nucleus upon elicitation, where it might regulate the expression of inducible defense genes. SIPK (salicylic-acid induced kinase) was shown to be activated post-translationally by different fungal elicitors from *P. parasitica* and *P. cryptogea* in tobacco suspension cells as well as by SA (Zhang and Klessig 1998). What direct or indirect role these related MAP kinases may play in ethylene signal transduction remains to be seen. In this respect it is of interest that signaling pathways can interchange MAP kinases without loss of signal specificity. This can be achieved by sequestering of the kinase components in pathway-specific signal modules as has been shown for yeast (reviewed in Madhani and Fink, 1998).

## 3.6
## EIN3: A Nuclear-Localized Component of Ethylene Signaling

Double mutants of *ein3/ctr1* show an ethylene insensitive phenotype, thus positioning *EIN3* downstream from *CTR1*. EIN3 was isolated by T-DNA tagging and found to be part of a multigene family called EIL (EIN3-LIKE, Chao et al. 1997). It is likely that EIN3 acts as a positive regulator of the ethylene pathway, because overexpression of wild-type EIN3 conferred a constitutive ethylene phenotype that is insensitive to the addition of inhibitors of ethylene sensing. Interestingly, *EIL* genes can also complement *ein2* mutants when regulated by the CaMV 35S viral promoter. EIN3 and EIL1 show interesting structural features including short stretches of positively charged amino acids that can serve as nuclear localization signals. Indeed EIN3-GUS reporter gene fusions are localized in the nucleus. Additional stretches of acidic, glutamine and proline-rich regions in EIN3 may serve as transcriptional activation domains. In this respect, it is interesting that fusions of EIN3 to the GAL4 DNA-binding domain could activate transcription in yeast (Chao et al. 1997). Whether or not transcriptional regulation is the actual function of the EIN- and EIL-class molecules remains to be demonstrated.

## 3.7
## PK12: A Kinase Exhibiting Ethylene-Dependent Activity

Direct biochemical approaches for isolating components of signal-transduction pathways can yield information on pleitropic components of the pathway that may be difficult to screen for by genetic means. A differential reverse transcriptase PCR approach revealed an ethylene-induced cDNA clone, encoding a protein kinase (Sessa et al. 1996). Elevated levels of the PK12 transcript were detected in leaves after ethylene treatment and constitutively in the flower abscission zone. The abscission zone is known for its high sensitivity to ethylene-mediated processes and is a tissue in which other ethylene-induced genes are present constitutively (Eyal et al. 1993). PK12 is a member of the novel kinase family containing the amino acid signature LAMMER in its C-terminus. Similar motifs were recently identified in mammals, flies and yeast. The decoded sequence contains a nuclear localization site (Hicks et al. 1995) and PK12-GUS fusion proteins accumulate in the nucleus in a manner similar to other members of this family (Savaldi-Goldstein, S and Fluhr, R, unpubl.). PK12 exhibits rapid and transient ethylene-dependent activation of kinase activity. The kinase activity of immunoprecipitated PK12 on the heterologous substrate MBP peaks within 5–10 min of ethylene application and then returns to a low basal level within 30 min. Transient stimulation is a hallmark of receptor mediated responses. It is associated with the phenomenon of desensitization and has been shown for other plant responses such as desensitization of the cellular reaction towards chitin fragments (Felix et al. 1998).

PK12 was analyzed biochemically in vitro. The recombinant PK12 has a dual-specificity activity, phosphorylating both tyrosine and serine and threonine residues (Sessa et al. 1996). The conserved LAMMER domain was found to be crucial for the activity. Amino acid substitution at the conserved LAMMER motif, in which the three amino-acids MME to RAQ respectively, caused a complete abrogation of PK12 activity. By contrast, a mutation in the invariant lysine, which abolishes the activity of most protein kinases, had only a reduced effect (Savaldi-Goldstein, S and Fluhr, R. unpubl.). The predicted location of the LAMMER motif, and its requirement for PK12 kinase activity, suggests that it might be responsible for substrate recognition.

Using the two-hybrid system, the mammalian Clk/Sty homologue of the LAMMER family, has been shown to bind SR splicing factors and to both phosphorylate and regulate their intracellular distribution in vivo (Colwill et al. 1996). SR splicing factors are proposed to participate in exon intron-bridging interactions, thus mediating commitment between pairs of splice sites (Chabot 1996). This group of proteins contain sequences of alternating serine and arginine, defined as RS domain, which may be involved in protein-protein interaction. SR splicing factor has been recently purified from plants. However, compared with the animal SR proteins which are conserved in size and number, they include many variants (Lopato et al. 1996). Interestingly, differential RNA display carried out on tobacco cell suspension elicited with cryptogein, led to the isolation of a cDNA homologue to the human transformer-2-like SR protein (Petitot et al. 1997). The corresponding transcript accumulated transiently in the cells within 3 h of cryptogein elicitation. The involvement of SR proteins in a pathogenesis response and the regulation of PK12 activity by ethylene necessitates further analysis possible interactions between these proteins. The activation characteristics of PK12 suggest its involvement in the ethylene signal transduction pathway, possibly coordinating general transcriptional or splicing mechanisms, as suggested for other members of the LAMMER family (Yun et al. 1994; Colwill et al, 1996).

## 3.8
### EREBPs: Transcription Factors That Bind cis-Regulatory Sequences

A subset of pathogenesis-related proteins accumulates to high levels in the presence of stress-induced ethylene. Comparison of the *cis*-regulatory sequences of the ethylene-sensitive genes revealed an 11 bp sequence, referred to as the GCC box, which was essential for ethylene-dependent transgene expression (Eyal et al. 1993; Vogeli-Lange et al. 1994; Sessa et al. 1995). Using a lambda phage expression library, genes encoding proteins that bind to GCC motifs were isolated (Ohme-Takagi and Shinshi 1995). The polypeptides were designated ethylene-responsive element binding proteins (EREBPS). A small region of homology within the EREBP contains the DNA binding site and shows significant similarity to the APETALA2 floral homeotic protein. Mutations in the homologous domains of the floral protein result in floral-malformation phenotypes, suggesting that the DNA binding region is necessary for gene function (Ecker 1995). Significantly, EREBP transcripts themselves were induced from low or undetectable basal levels to higher levels in the presence of ethylene.

## 4
# Dynamics of Ethylene Perception and Signal Transduction

## 4.1
### Perception of Ethylene and Phospho-Transfer

Although ETR1 kinase activity has not been experimentally established, the precisely conserved residues that can participate in the phosphotransfer make its biological activity as a kinase likely. Inspection of the behavior of other two-component systems can illustrate how sensory information is processed. The cyanobacterial two-component phytochrome, Cph1, displays a spectral-sensitive conformation change. The $P_r$ form of this molecule is abundant in far-red light and rapidly undergoes autophosphorylation. The phosphorylated Cph1 is then capable of supporting phosphotransfer to Rcp1, a response regulator molecule. Conversely, the $P_{fr}$ form is abundant in red light and autophosphorylates at a much lower rate (Yeh et al. 1997).

Another useful paradigm for ethylene perception is the hybrid histidine kinase-regulator Sln1p, which functions in the yeast high osmolarity (HOG) pathway. Under conditions of low osmolarity, Sln1p autophosphorylates on His576, which is subsequently shifted to Asp1444 (Posas et al. 1996). By analogy with the yeast osmosensing system involving Sln1p, the *ETR1* gene product in plant cells may be in the active state in the absence of ethylene. When ETR1-like genes were mutated to obtain loss of function alleles no apparent phenotype was evident. However, combining the loss-of-function mutants of the four different ethylene receptor homologues into one plant resulted in an ethylene constitutive response (Hua et al. 1998a). This indicates that the proteins negatively regulate the ethylene response pathway. In the absence of ethylene, ongoing phospho-relay activity would activate downstream CTR1 and thus effectively repress ethylene responses. In this scenario, the mutations in the membrane-spanning region would lock ETR1 into a constitutively active state and could explain the dominant nature of ETR1 mutations. This is consistent with the finding that truncation of the amino portion of the transmitter leads to constitutive kinase activity in bacterial two-component sensors (Parkinson and Kofoid 1992). However, amino acid replacements at the conserved ETR1 kinase sites that would most certainly destroy kinase activity, fail to attenuate the dominant mutant phenotype (Chang and Meyerowitz 1995); this finding is evidence against a constitutive activation model. One possibility is that transmitters can function as dimers and display cross-phosphorylation activity as suggested for SLN1p. Thus, ETR1 or ERS may interact and exchange activation states, enabling mutants with constitutive expression to effectively 'poison' the system.

## 4.2
## Linking Phospho-Relay and Kinase Cascades

The transmission of extracellular signals generally involves sequential activation of MAP (mitogen-activated) kinases. The HOG MAP kinase pathway is responsible for stimulating the yeast physiological response to the osmotic environment, and receives input from the sensor kinase SLN1p (Fig 1). SLN1p works in conjunction with two downstream molecules to transfer phosphate sequentially to His64 of Ypd1p and finally to Asp554 of Ssk1p. Phosphorylated Ssk1p is incapable of activating SSK2p, the first in the hierarchy of MAP kinases (MAPKKK). At high osmolarity, SLN1p autophosphorylation is inactivated and subsequently Ssk1p is dephosphorylated and the HOG pathway is activated. Osmotic sensing in yeast is an example of an emerging pattern of phospho-relay in which phosphate is sequentially transferred between histidine and aspartate-containing modules, which finally modulate a MAPK cascade. The phospho-relay mechanism does not amplify input signals. Instead, each module serves as a potential regulatory control point for integrating complex physiological processes. Molecules that probably bridge between histidine phospho-relay systems and MAP kinase cascades are the recently described *Arabidopsis* response regulators (ARR; Imamura et al. 1998). Alternatively, ETR1 may directly regulate CTR1 activity. Indeed, two-hybrid analysis carried out in yeast established a direct interaction between ETR1 and CTR1 (Clark et al. 1998). The predicted receiver domain of ETR1 and ERS was shown to specifically interact with the CTR1 amino-terminal domain. Furthermore, the interacting portion of CTR1 corresponds to the regulatory region of Raf, which is responsible for its association with Ras and 14-3-3 proteins. Nevertheless, the two-hybrid system by itself is not sufficient to determine the situation in vivo. Therefore, the relevance of the above interactions has to be verified in planta. It remains to be seen whether phospho-relay takes place via independent response regulator modules as the SLN1 paradigm predicts, or by direct ETR1/CTR1 interaction.

CTR1 is analogous to Raf-1, which can function at a hierarchical MAPKKK level. The negative mode by which CTR1 asserts regulation does not make it directly homologous to the Ssk2p-type of MAPKKK present in yeast osmosensing. CTR1 could negatively regulate a MAPKK kinase directly. In the absence of ethylene, ETR1 would activate CTR1, which would impose signal quiescence by phosphorylation of a putative MAPKK (Fig. 1). Unfortunately, candidates for MAP kinases have yet to be resolved by genetic means among the loci isolated. It is possible that functional redundancies of the MAP kinases make their isolation difficult. Alternatively, despite the apparent dispensability of the ethylene pathway, mutations at MAPK loci might be lethal due to interchangeability of the components with other signaling pathways (Madhani and Fink 1998)

The finding that *EIN3* encodes a nuclear-localized component of the transcription pathway is an important link which connects the phospho-relay and subsequent kinase cascades to gene activation. The question of what mediates signal transfer to the nuclear compartment and whether EIN3 or similar molecules interact directly with EREBP-like or LAMMER kinases remains to be addressed. The structural and functional similarity of prokaryote and eukaryotic signaling pathways will undoubtedly continue to provide a rich source of useful, sometimes universal, paradigms of cellular perception.

## 4.3
## Redundancy of Components and Harmonizing the Ethylene Response

The existence of multigene families in signaling components of the ethylene pathway raises the question of redundancy. In this respect, members of the gene families display differential but overlapping expression patterns. For example, ERS1, ERS2 and ETR2 are induced by ethylene at the transcriptional level, whereas ETR1 and EIN4 are not (Hua et al. 1998b). In the case of ETR2, differential enrichment of the transcript was found in petals and ovules. Toma-to plants express a constitutive low level of an ETR1-like component in all tis-sues (Zhou et al. 1996) and an ERS-type receptor (Nr) which is ethylene indu-cible in ripening-competent fruit and flower abscission zones (Payton et al. 1996). Notwithstanding, loss-of-function mutants of any single receptor of *Arabidopsis* does not lead to an obvious phenotype suggesting redundancy by the other intact family members (Hua et al. 1998a). This raises the question of how seemingly redundant genes are maintained in the genome and suggests that environmental conditions exist that necessitate the expression of each gene. Conceivably, tissue-specific requirements or compensation for different ethylene-sensitivity in the receptors may dictate this division of labor. Howev-er, the fact that the *Nr* mutation, which occurs in an ethylene inducible ERS-like component, can confer dominant ethylene-insensitivity throughout the plant, argues for ongoing complex interactions between these components.

A further complication may be the presence of alternate splicing as has been detected for *ETR2* which would produce components with or without the C-terminal receiver domain. The importance of the C-terminal receiver domain can be illustrated in the hybrid-type ArcB bacterial sensor responsible for an-aerobic sensing. Deletion of the response-regulator domain from the hybrid-type histidine kinase ArcB alters the strength of regulation, but does not pre-vent anaerobic regulation (Iuchi and Weiner 1996).

Another special feature of the ethylene pathway is the sensitivity of tran-script level to the presence of ethylene. Elevated ETR1-like transcripts could indicate feed-back repression of the ethylene response. For example, if ethylene induced *ERS1*, *ERS2* and *ETR2* produce receptor with lower binding constants for ethylene the elevated expression could lead to attenuation of the response. Other components in the pathway show modulation by ethylene. PK12 tran-script is induced by ethylene and is also present in elevated amounts in ethy-lene-sensitive tissue, yet the constitutive levels of PK12 also show a marked ethylene-dependent increase in their phosphorylating activity (Sessa et al. 1996). The ethylene-responsive element binding transcript level is elevated by ethylene as well (Ohme-Takagi and Shinshi 1995). In all cases it is unknown if the elevated transcript levels results in more active protein or compensation for accelerated turnover of the components.

*EIN3* transcripts, in a fashion similar to *ETR1* and *CTR1* transcripts, do not show marked sensitivity to ethylene. However, the complexity of signal-path-way control at this juncture is illustrated by the result that *EIL* genes can com-plement the *ein3* mutation when overexpressed in transgenic plants. Yet *EIL* homologues cannot normally substitute for the *EIN3* function, as *ein3*-mutant phenotypes are readily detectable in genetic screens. Once again, is the expres-sion of these genes temporally and spatially separated in the plant, or does gross overexpression lead to exhibition of latent but non-physiological func-tions? Other examples of seemingly redundant functions and ethylene-indu-cible signaling components are the ethylene *trans*-acting factors (Ohme-Takagi and Shinshi 1995; Zhou et al. 1997). Although initially discovered as binding to a specific ethylene-box-containing sequence in pathogenesis-related proteins, their in vivo specificity may show marked differential affinities. Obviously, detailed biochemical analysis of the *trans*-acting factors will be necessary to establish the manner in which they function. It is evident that the maintenance of seemingly redundant components in the pathway ensures that the cellular response culminates in a well-coordinated synchronously robust tissue-level outcome.

The identification of the gene products ETR1, ARR and CTR1, as histidine, response regulators and Raf-like kinases, respectively, implies that the activity of cellular kinases and complementary phosphatases play an important role in ethylene responses. For fuller understanding of this signal transduction, the putative activity of these genes as well as essential ethylene-linked mitogen-activated protein kinases remains to be demonstrated.

*Acknowledgements.* This work was supported by grants from the Israel Science Foundation administered by the Israel Academy of Sciences and Humanities.

# References

Bleecker AB, Estelle MA, Somerville C, Kende H (1988) Insensitivity to ethylene conferred by a dominant mutation in *Arabidopsis thaliana,* Science 241:1086–1089

Brandstatter I, Kieber JJ (1998) Two genes with similarity to bacterial response regulators are rapidly and specifically induced by cytokinin in *Arabidopsis*. Plant Cell 10:1009–1019

Burg SP, Burg EA (1966) Molecular requirements for the biological activity of ethylene. Plant Physiol 42:144–152

Chabot B (1996) Directing alternative splicing: cast and scenarios. Trends Genet 12:472–478

Chang C, Meyerowitz E (1995) The ethylene hormone response in *Arabidopsis:* a eukaryotic two-component signaling system. Proc Natl Acad Sci USA 92:4129–4133

Chang C, Kwok SF, Bleecker AB, Meyerowitz EM (1993) *Arabidopsis* ethylene-response gene *ETR1:* similarity of product to two-component regulators. Science 262:539–545

Chao Q, Rothenberg M, Solano R, Roman G, Terzaghi W, Ecker JR (1997) Activation of the ethylene gas response pathway in *Arabidopsis* by the nuclear protein ethylene-insensitive 3 and related proteins. Cell 89:1133–1144

Clark KL, Larsen PB, Wang XX, Chang C (1998) Association of the *Arabidopsis* CTR1 Raf-like kinase with the ETR1 and ERS ethylene receptors. Proc Natl Acad Sci USA 95:5401–5406

Colwill K, Oawsib T, Andrews B, Prasad J, Manley JL, Bell JC, Duncan PI (1996) The Clk/Sty protein kinase phosphorylates SR splicing factors and regulates their intranuclear distribution. EMBO J 15:265–275

Dietrich A, Mayer JE, Hahlbrock K (1990) Fungal elicitor triggers rapid, transient, and specific protein phosphorylation in parsley cell suspension cultures. J Biol Chem 265:6360–6368

Ecker JR (1995) The ethylene signal transduction pathway in plants. Science 268:667–675

Eyal Y, Meller Y, Lev-Yadun S, Fluhr R (1993) A basic-type PR-1 promoter directs ethylene responsiveness, vascular and abscission zone-specific expression. Plant J 4:225–234

Felix G, Baureithel K, Boller T (1998) Desensitization of the perception system for chitin fragments in tomato cells. Plant Physiol 117:643–650

Fluhr R, Mattoo AK (1996) Ethylene biosynthesis and perception. Crit Rev Plant Sci 15:479–523

Goeschl JD, Rappaport L, Pratt HK (1966) Ethylene as a factor regulating the growth of pea epicotyls subjected to physical stress. Plant Physiol 41:877–884

Grosskopf DG, Felix G, Boller T (1990) K-252a inhibits the response of tomato cells to fungal elicitors in vivo and their microsomal protein kinase in vitro. FEBS Lett 275:177–180

Gupta WN, Roberts TM (1994) Signal transduction pathways involving the Raf proto-oncogene.
Cancer Metastasis Rev 13:105–116

Guzman P, Ecker JR (1990) Exploiting the triple response of Arabidopsis to identify ethylene-related mutants. Plant Cell 2:513–523

Heldin CH (1995) Dimerization of cell surface receptors in signal transduction. Cell 80:213–223

Hicks GR, Smith HMS, Shieh M, Raikhel NV (1995) Three classes of nuclear import signals bind
to plant nuclei. Plant Physiol. 107:1055–1058

Hirt H (1997) Multiple roles of MAP kinases in plant signal transduction. Trends Plant Sci 2:11–15

Hua J, Caren C, Sun Q, Meyerowitz EM (1995) Ethylene insensitivity conferred by Arabidopsis
ERS gene. Science 269:1712–1714

Hua J and Meyerowitz EM (1998a) Ethylene responses are negatively regulated by a receptor gene
family in Arabidopsis thaliana. Cell 94:261–271

Hua J, Sakai j, Nourizadeh S, Chen QG, Bleecker AB, Ecker JR, Meyerowitz EM (1998b) EIN4 and
ERS2 are members of the putative ethylene receptor gene family in Arabidopsis. Plant Cell
10:1321–1332

Imamura A, Hanaki N, Umeda H, Nakamura A, Suzuki T, Ueguchi C, Mizuno T (1998) Response
regulators implicated in His-to-Asp phosphotransfer signaling in Arabidopsis. Proc Natl
Acad Sci USA 95:2691–2696

Iuchi S, Weiner L (1996) Cellular and molecular physiology of Escherichia coli in the adaptation
to aerobic environments. J Biochem 120:1055–1063

Kakimoto T (1996) CKI1, a histidine kinase homolog implicated in cytokinin signal transduction. Science 274:982–985

Kieber J, Rothenberg M, Roman G, Feldmann K, Ecker J (1993) CTR1, a negative regulator of the
ethylene response pathway in Arabidopsis, encodes a member of the Raf family of protein
kinases. Cell 72:427–441

Kieber JJ (1997) The ethylene response pathway in Arabidopsis. Annu Rev Plant Physiol Plant
Mol Biol 48:277–296

Kwak SH, Lee SH (1997) The requirements for Ca2+ protein phosphorylation, and dephosphorylation for ethylene signal transduction in Pisum sativum L. Plant Cell Physiol 38:1142–1149

Lanahan MB, Yen HC, Giovannoni JJ, Klee HJ (1994) The never ripe mutation blocks ethylene
perception in tomato. Plant Cell 6:521–530

Ligterink W, Kroj T, Nieden UZ, Hirt H, Scheel D (1997) Receptor-mediated activation of a MAP
kinase in pathogen defense of plants. Science 276:2054–2057

Lopato S, Waigmann E, Barta A (1996) Characterization of a novel arginine/serine-rich splicing
factor in Arabidopsis. Plant Cell 8:2255–2264

Lynch BAU, Koshland DE Jr (1991) Disulfide cross-linking studies of the transmembrane regions
of the aspartate sensory receptor of Escherichia coli. Proc Natl Acad Sci USA 88:10402–10406

Machida Y, Nishihama R, Kitakura S (1997) Progress in studies of plant homologs of mitogen-activated protein (MAP) kinase and potential upstream components in kinase cascades. Crit
Rev Plant Sci 16:481–496

Madhani HD, Fink GR (1998) The riddle of MAP kinase signaling specificity Trends Genet 14:
151–155

Mattoo AK, Suttle CS (1991) The Plant Hormone Ethylene. CRC Press, Boca Raton, Florida

McGrath RB, Ecker JR (1998) Ethylene signaling in Arabidopsis: events from the membrane to
the nucleus. Plant Physiol Biochem 36:103–113

Milligan AU, Koshland DE Jr (1988) Site-directed cross-linking. Establishing the dimeric structure of the aspartate receptor of bacterial chemotaxis. J Biol Chem 263:6268–6275

Mizoguchi T, Ichimura K, Shinozaki K (1997) Environmental stress response in plants: the role
of mitogen-activated protein kinases. Trends Biotechnol 15:15–19

O'Donnell PJ, Calvert, C, Atzorn R, Wasternack C, Leyser HMO, Bowles D J (1996) Ethylene as a signal mediating the wound response of tomato plants. Science 274:1914–1917

Ohme-Takagi M, Shinshi H (1995) Ethylene-inducible DNA binding proteins that interact with an ethylene-responsive element. Plant Cell 7:173–182

Pan AU, Charles T, Jin S, Wu ZL, Nester EWSO (1993) Preformed dimeric state of the sensor protein VirA is involved in plant-*Agrobacterium* signal transduction. Proc Natl Acad Sci USA 90:9939–9943

Parkinson JS, Kofoid EC (1992) Communication modules in bacterial signaling proteins. Annu Rev Genet 26:71–112

Payton S, Fray G, Brown S, Grierson D (1996) Ethylene receptor expression is regulated during fruit ripening, flower senescence and abscission. Plant Mol Biol 31:1227–1231

Petitot AS, Blein JP, Pugin A, Suty L (1997) Cloning of two plant cDNAs encoding a beta-type proteasome subunit and a transformer-2-like SR-related protein: early induction of the corresponding genes in tobacco cells treated with cryptogein. Plant Mol Biol 35:261–269

Posas F, Susannah M, Wurgler-Murphy, Maeda T, Witten, EA, Thai TC, Saito H (1996) Yeast HOG1 MAP kinase cascade is regulated by a multistep phosphorelay mechanism in the SLN1-YPD1-SSK1 "two-component" osmosensor. Cell 86:865–875

Raz V, Fluhr R (1992) Calcium requirement for ethylene-dependent responses. Plant Cell 4:1123–1130

Raz V, Fluhr R (1993) Ethylene responsiveness is transduced via phosphorylation events in plants. Plant Cell 5:523–530

Rodriguez FI, Esch JJ, Hall Ae, Binder BM, Schaller GE, Bleecker AB (1999) A copper cofactor for the ethylene receptor ETR1 from *Arabidopsis*. Science 283:996–998

Roman G, Lubarsky B, Kieber JJ, Rothenberg M, Ecker JR (1995) Genetic analysis of ethylene signal transduction in *Arabidopsis thalia*. Genetics 139:1393–1409

Sakai H, Hua J, Chen QHG, Chang CR, Medrano LJ, Bleecker AB, Meyerowitz EM (1998) ETR2 is an ETR1-like gene involved in ethylene signaling in *Arabidopsis*. Proc Natl Acad Sci USA 95:5812–5817

Schaller GE, Bleecker AB (1995) Ethylene-binding sites generated in yeast expressing the *Arabidopsis* ETR1 gene. Science 270:1809–1811

Schaller GE, Ladd AN, Lanahan MB, Spanbauer JM, Bleecker AB (1995) The ethylene response mediator ETR1 from *Arabidopsis* forms disulfide-linked dimer. J Biol Chem 270:12526–12530

Sessa G, Meller, Y, Fluhr R (1995) A GCC element and a G-box motif participate in ethylene-induced expression of the PRB-1b gene. Plant Mol Biol 28:145–153

Sessa G, Raz V, Savaldi S, Fluhr R (1996) PK12, a plant dual-specificity protein kinase of the LAMMER family, is regulated by the hormone ethylene. Plant Cell 8:2223–2234

Taylor JE, Grosskopf DG, McGaw BA, Horgan R, Scott IM (1988) Apical localization of 1-aminocyclopropane-1-carboxylic acid and its conversion to ethylene in etiolated pea seedlings. Planta 174:112–114

Vogeli-Lange R, Frundt C, Hart CM, Nagy F, Meins F Jr (1994) Developmental hormonal and pathogenesis-related regulation of the tobacco class 1b-1,3-glucanase B promoter. Plant Mol Biol 25:299–311

Wilkinson JQ, Lanahan MB, Yen HC, Giovannoni JJ, Klee HJ (1995) An ethylene-inducible component of signal transduction encoded by *Never-ripe*. Science 270:1807–1809

Wu Y, Kuzma J, Marechal E, Graeff R, Lee HC, Foster R, Chua NH (1997) Abscisic acid signaling through cyclic ADP-Ribose in plants. Science 278:2126–2130

Yeh KC, Wu SH, Murphy JT, Lagarias JC (1997) A cynobacterial phytochrome two-component light sensory system. Science 277:1505–1508

Yun B, Farkas R, Lee K, Rabinow L (1994) The *Doa* locus encodes a member of a new protein kinase family and is essential for eye and embryonic development in *Drosophila malanogaster*. Genes Dev 8:1160–1173

Zhang S, Klessig DF (1998) The tobacco wounding-activated mitogen-activated protein kinase is encoded by SIPK. Proc Natl Acad Sci USA 95:7225–7230

Zhang S, Du H, Klessig DF (1998) Activation of the tobacco SIP kinase by both a cell wall-derived carbohydrate elicitor and purified proteinaceous elicitins from *Phytophthora* spp. Plant Cell 10:435–450

Zhou D, Kalaitzis P, Mattoo AK, Tucker ML (1996) The mRNA for an ETR1 homologue is constitutively expressed in vegetative and reproductive tissues. Plant Mol Biol 30:1331–1338

Zhou J, Tang X, Martin GB (1997) The Pto kinase conferring resistance to tomato bacterial speck disease interacts with proteins that bind a *cis*-element of pathogenesis-related genes. EMBO J 16:3207–3218

# Subject Index

ABI1/2   98, 135, 138, 139

Abiotic stresses *see also* Cold stress, Dehydration, Drought stress, Heat stress, Hyperosmotic stress, Hypoosmotic stress, Mechanical stress, Osmotic stress, Oxidative stress, Salt stress, and Touch stress   1, 5, 8, 131, 133, 139

Abscisic acid (ABA)   8, 20, 45, 54, 55, 97, 98, 131–141, 147

ACC (1-aminocyclopropane-1-carboxylic acid)   146

Actin   20, 42, 44

Activation loop   14, 17, 24

*Agrobacterium*   150

Aldose reductase   133

Aleurone cells   8, 20, 45, 79, 98, 132–141

Alfalfa   4, 5, 7, 14, 33, 54, 55, 80, 99, 102, 103

Alkalization   47, 69–72

Amiprophosmethyl (APM)   103

Amphotericin B   90

Anaphase   7, 20, 102–104, 110, 111

ANP1-3   32, 123, 124

Anthracene-9-carboxylate (A9C)   89

APETALA2   154

Aphidicolin   102, 125, 126

*Arabidopsis*   4, 5, 8, 14, 19–21, 31–35, 45, 54, 67, 98, 105, 123, 134, 135, 139, 140, 146–151, 156, 157

ArcB   157

Archidonic acid   78

AsPK9   15, 18, 79, 98

AsPK10   98

ATATPK1   98

ATHK1   35

ATMEKKs
  ATMEKK1   34
  ATMEKK1-4   32

ATMKKs
  ATMKK2   34
  ATMKK2-5   32

ATMPKs
  ATMPK1   15, 19, 24, 32

ATMPK2/5/7   15, 32

ATMPK3   15, 17–19, 24, 32–34, 45, 57, 79, 88, 138

ATMPK4   15, 19, 32, 34

ATMPK6   15, 18, 32, 34, 80

ATMPK8/9   15, 32, 34

ATP-binding site   13

ATPP2Ci   139

Aux1   98

Auxin   8, 20, 97–100, 126

Avirulence-genes   66, 72, 85, 86
  Avr9   72, 74–76

Avocado   146

Bacterial elicitor *see also* harpin   72

Banana   146

Barley   8, 20, 45, 98, 132, 135, 137, 140, 141

BCK1   32, 34, 120, 121, 124

Brassica   41

Broad bean   140

$Ca^{2+}$ changes   41, 42, 45, 88, 106, 134

Calreticulin   47

Calyculin A   67, 68

Cantharadin   67

Carborundum   76

Carrot   132

CDC25 phosphatase   101

CDC7   104

CDK1   101

CDPK   42

Cell cycle   1, 7, 8, 19, 20, 53, 97–102, 107, 125–127, 133

Cell division   95, 97–103, 105, 107, 110, 111, 120, 125

Cell plate formation   105, 110, 128

Cell plate   7, 20, 105

Cell wall   3, 20, 45, 86, 119, 121

Cell wall-derived carbohydrate (CWD) elicitors   69–72, 75–79

CENP-E   100, 103

Centrosomes   100, 108, 109

Cf-9   72, 74–76

Chalcone synthase (CHS)   67
Checkpoint control   100–103, 110
CheY   151
Chitinase   146
Chitosan   6
Chromosome alignment   100–103, 110
CKI1   35, 151
CL100   100
*Cladosporium fulvum*   72
*Clamydomonas*   140
Classification   3, 14, 17–21, 31, 34
Cold stress   5, 19, 32–35, 43, 45, 46, 57, 133,
   135, 138, 140
*Commelina*   134
Cortical microtubules   105–108
Cotyledons   124–126
Cph1   154, 155
Cryptogein   69, 70, 154
CTR1   8, 20, 32, 35, 146, 147, 151, 152,
   155–158
Cutting   *see* wounding
Cyclin-dependent kinase (CDKs) *see also*
   CDK1   11, 12, 97, 101, 107–111, 120, 126
Cyclins *see also* $G_1$, $G_2$ and M-phase cyclins
   120
Cytokinesis   8, 100, 102–105, 119
Cytokinin   97, 98, 151
Cytoskeleton   2, 7, 20, 42, 47, 95, 102, 105,
   106, 110

Def1   6
Defense gene activation   88, 89, 91
Defense related genes   53, 57
Defense response genes *see also* Chalcone
   synthase (CHS), Phenylalanine ammonia
   lyase (PAL), PR genes, and Proteinase
   inhibitors   5, 66–69, 78.152
Defense response        66–69, 73–76, 78,
   86–90
Dehydration   40–44, 132, 133
Differential splicing   124, 154
Differentiation   3, 46, 47, 67, 79, 99, 102, 132,
   133, 151
Diphenylene iodonium (DPI)   88, 91
Disease resistance   66, 71, 76, 77
Division plane   104, 105, 107
Drought stress *see also* dehydartion   8, 19,
   33, 35, 43–46, 57, 133, 135, 138
Dynamin   105
Dynein   109

E2F   97, 99
EF-hand   139
Eg5   109, 111

EIF-4A   43
EIL gene family   152, 157
   EIL1   152
EIN gene family   146, 147, 151, 152
   EIN2   147, 152
   EIN3   152, 156, 157
   EIN4   149, 151, 157
Electrical current   54, 55, 58, 106
Elicitins *see also* Cryptogein, and Parasiticein
   69–72, 74, 76, 79
Elicitors *see also* Bacterial elicitors, Fungal
   elicitors, and Microbial elicitors   6, 19, 34,
   46, 47, 57, 59, 66, 69–72, 75, 77, 78, 85, 86,
   152
   non-race specific   66, 72, 76, 86
   race-specific   66
Embryos   124–126, 138
Environmental stresses *see also* Abiotic
   stress, Cold stress, Dehydration, Drought
   stress, Heat stress, Hyperosmotic stress,
   Hypoosmotic stress, Insect attack,
   Mechanical stress, Osmotic stress,
   Oxidative stress, Salt stress, Touch stress,
   and Wounding   3, 4, 8, 32, 33, 36, 40, 67,
   79, 132
Era1–1   134
EREBPs        154, 156
ERK1/2   3, 7, 12, 13, 16–18, 45, 78, 79, 96, 97,
   100–105, 110, 135
ERMK   7, 15, 33, 57, 79, 81, 152
ERS   8, 149, 155–157
   ERS1/2   149, 151, 156, 157
*Erwin amylovora*   6, 72
ESTs   4, 20, 21, 24
Ethylene   8, 32, 53, 97, 98, 145–150, 152–158
ETR1   8, 9, 35, 146–158
ETR2   149, 156, 157

Farnesyl transferase   134, 141
Fertilization   145
Flavonols   41
Fruit ripening   145, 146, 157
Fungal elicitors *see also* Cell wall-derived
   carbohydrate (CWD) elicitors, Cryptogein,
   Elicitins, Parasiticein, Peptide elicitors   67,
   69, 71, 74, 76–79, 147
FUS3   3, 31, 102, 121

G/HBF-1   67
$G_1$
   Arrest   99
   Control   95, 98–100
   Cyclins   99
      Cyclin D3   98, 99

Cyclin D1   97
Phase   99, 100, 102
$G_1/S$
   Boundary   102
   Transition   99
$G_2$
   Arrest   101
   Control   102
   Cyclins   107
   Phase   100, 102, 107
$G_2/M$ transition   100
$Gd^{3+}$   88
Gene for gene hypothesis   72, 75, 85
Germination   5, 19, 40–44, 132–134
Giberellic acid   8, 20, 79, 97–99, 132, 137, 139
Glycogen synthase kinase (GSK)   11, 97
Glycoproteins   7, 86
Guard cells   134, 138–140

H7   146
HAPK   72
Harpin   6, 72, 74, 76, 78
Heat stress   3, 30, 33, 46, 78, 121, 138
Herbivore   53
Histidine kinase   33, 35, 98, 148–151, 155, 157
Histidine phosphorylation   9, 150, 155, 156
HOG1   3, 9, 31, 33, 35, 45, 46, 155
Hormones see also Abscisic acid, Auxin, Cytokinin, Ethylene, Giberellic acid, and Plant hormones   3, 46
Host-recognition   66, 85
Hydration see rehydration
Hydraulic pressure   58
Hyperosmotic stress   3, 33, 45, 78, 121, 133
Hypersensitive cell death   70, 71
Hypersensitive respons (HR)   66, 67, 69, 73, 74
Hypocotyl   124, 146
Hypoosmotic stress   3, 33, 45, 46

IBC6/7   151
ICK1   98, 99
Incompatible interaction   66
Insect attack   53, 78
Integrin   47
Intercellular pH   134, 139
Interphase   103, 106–110
Ion fluxes   66, 87–91, 134

Jasmonate synthesis   58, 59, 78, 88
Jasmonate-inducible genes   58
Jasmonic acid   6, 19, 41, 53–55, 58, 59, 78
JNK   3, 30, 33, 45, 78, 79

K252a   67, 102, 107, 108, 140, 146, 147
KAP1/2   67
Keule   105
Kinase Domains   3, 12, 14, 68, 123
Kinesin   100, 103, 111, 127
Kinetochores   100, 108–110
Knolle   105
$KO_2$   88, 91
KSS1   3, 31, 121

$La^{3+}$   88
LAMMER motif   153, 156
LePRK1/2   41
Leucine rich repeats (LRR)   41, 72
Leucotrienes   78
Light   106, 155
Linolenic acid   78
L-malate   134
Low temperature stress see cold stress

Maize   20, 42, 108
MAPK inactivation   6, 12, 55, 75, 98, 139
MAPK-like activity   6, 8, 9, 20, 33, 34, 54, 57, 69–72, 79, 135, 138
Mechanical stress   4, 5, 19, 33, 74, 78, 106
Mechanoreceptor   46, 47
Meiosis   101
MEK1   32, 34
Metaphase   100–103, 110
Microbial elicitors   6, 67
Microtubule motors   105, 109, 111, 127, 128
Microtubule-associated proteins (MAPs)   106, 108, 109
Microtubules see also cortical microtubules   101–111, 119, 120, 128
Mitogens   3, 8
Mitosis   7, 20, 39, 40, 43, 100–104, 107, 109, 110
Mitotic spindle   105, 108, 109, 110
MMKs
   MMK1   15, 19, 45, 46, 55, 79, 81, 99, 102–104, 140
   MMK2   15, 20, 55, 81, 140
   MMK3   7, 15, 20, 24, 81, 102–105
   MMK4   15, 19, 33, 34, 45, 46, 55–58, 79–81, 88, 138
MP2C   6, 139
M-phase   128
   Cyclins   126, 128
MPK1   3, 20, 31, 33, 34, 45, 121
MPM2   109, 110
MYT1 kinase   101

N gene   47, 73, 74
NACK1/2   7, 127, 128
NADPH oxidase   7, 88, 91
NaHG   74
Necrosis   69, 72
Necrotic lesions   66, 146
Non-host recognition   85, 86
NPKs
    NPK1   7, 8, 20, 32, 98, 103, 108, 120–128
    NPK2   32
*Nr*-gene   149, 151, 15
NtFs
    NtF3   15, 43, 55
    NtF4   5, 15, 19, 43–46, 55, 56, 70, 80
    NtF6   7, 15, 20, 55, 102–105
Nuclear localization   2, 5, 7, 57, 78, 91, 97,
    100, 107, 152, 153, 156
Nucleolin   99
NuMA   109

Oat   8, 14, 20, 79, 133
Octadecanoid pathway   78
Okadaic acid   67, 68, 140, 147
Oocytes   101, 102
Osmotic stress *see also* Hyperosmotic stress,
    and Hypoosmotic stress   5, 33, 34, 47, 74,
    133, 155
Oxidative burst *see also* Reactive oxygen
    species (ROS)   7, 88–91
Oxidative stress   46

P38 kinase   3, 16, 17, 18, 30, 33, 45, 78
*Papaver rhoeas*   41, 42
Parasiticein   69, 70
Parsley   6, 14, 54, 67, 81, 86–89, 91
Pathogene response   20, 78, 86, 91, 145, 152,
    154
Pathogens   1, 5–7, 19, 47, 53, 66, 67, 72, 77,
    78, 85, 86, 131
PBF-1   68
PBS2   34, 35, 121
PDK1/2   97
Pea   14, 107, 140
*Penniseteumi*   132
Pep13 elicitor   86–91
Pep25 elicitor   81
Peptide elicitor *see also* Pep13, and Pep25
    elicitor   7
Petunia   14, 41
Phenyl arside oxide (PAO)   140
Phenylalanine ammonia lyase (PAL)   67,
    69–72
PhERK1   15, 43
Pheromone pathway   3, 31, 32, 121

Phospholipase A   78
Phospholipids   97
Phragmoplast   7, 103, 105, 110, 111, 128
Phragmoplastin   105
Phylogenetic analysis   18, 21, 31, 32, 79, 80
Phytoalexin synthesis   7, 67, 88, 89, 91
*Phytophthora*   6, 69, 72
    *cryptogea*   69, 152
    *infestans*   69, 71
    *parasitica*   69, 71, 152
    *sojae*   7, 81, 86
PI3K   97
PK12   153, 154, 157
PKA   120
PKB   97
PKC   31, 45, 68, 120, 121
Plant hormones *see also* Abscisic acid, Auxin,
    Cytokinin, Ethylene, and Giberellic acid
    1, 20, 106, 131, 132
Pollen   5, 19, 39–44, 46, 102
    Grains   39–44, 70
    Tube growth   40, 42–44
Polygalacturonic acid (PGA)   6
Potato   74
PR genes   5, 7, 67, 68, 73, 146, 147
    Acidic   58, 67
    Basic 53
    PR1   58, 68
    PR10   68
Preprophase band (PPB)   102, 105, 107, 108,
    110
Primordia   125
PRK1   41
Profilin   44
Proliferation   3, 8, 78, 99, 100, 119, 120
Prophase   100–103, 110
Propyzamide   102
Prostaglandins   78
Protein phosphatases
    PP1   147
    PP2A   109, 147
    PP2C *see also* MP2C, ATPP2Ci, and ABI1/2
        6, 98, 141
Proteinase inhibitors   5, 53, 67
    PI-II   58, 59
PsD5 (PsMAPK)   15, 45, 80
*Pseudomonas syringae*   72

Q gene   127

Raf kinase   8, 20, 32, 35, 96, 99–101, 127, 151,
    152, 156, 158
Ras   97–100, 134, 135, 140, 141
Rcp1   155

Reactive oxygen species (ROS)   47, 66, 69,
   88, 91
Rehydration   5, 19, 42–46, 70, 102
Resistance genes *see also* Cf-9, and N-gene
   66, 72, 76, 85, 86
Resistant plants   66, 85
Retinoblastoma protein (Rb)   97, 99, 120
RhoA   47
Rice   20
Roscovitine   102, 108, 110

SA-inducible genes   58, 68
Salicylic acid (SA)   19, 34, 46, 55–59, 67, 68,
   70–79, 152
Salt stress   5, 8, 19, 32, 35, 43, 45, 133, 135,
   138, 140
SAMK *see* MMK4
SAPK *see* JNK
Seed ripening   132
Senescence   145, 146
Signature sequence for MAPK   24, 44
SIPK   6, 15, 19, 34, 45–47, 55–60, 68–81, 152
SLN1   35, 46, 151, 155, 156
SLT2 *see* MPK1
SMK1   3, 31
Somatic cells   100, 102
Soybean   54, 78, 105
Spc1   46
S-Phase   99, 100, 107, 126
Spindle assembly   103, 108–110
Spindle pole bodies   104
Spore wall formation   3, 31
SR splicing factors   153, 154
Src   100,
SSK2 / SSK22   35, 121, 155, 156
Stathmin/Op18   109
Staurosporin   67, 71, 106–108
STE11   32, 104, 120, 121, 124, 139
Stigma   40, 41
Stomatal guard cells   133
Subfamily   3, 14, 15, 17–21, 24, 32, 79
Susceptibile plants   66, 85

*Synechocystis*   151
Systemic acquired resisitance (SAR)   67, 69,
   74
Systemic resistance   67
Systemic signaling   5, 6, 53, 57, 58, 74
Systemin   6, 54, 76

Tapetum   40
TAU   107
Taxol   102
Telophase   7, 20, 102–105
TMV   6, 46, 47, 67, 73–76
Tobacco   5–8, 14, 20, 32–34, 44–47, 54, 55,
   57, 67–76, 79, 81, 99, 102, 103, 109, 120, 125,
   140, 154
   Xanthi (nn)   74
   Xanthi nc (NN)   73, 74
Tomato   6, 41, 54, 57, 67, 72, 74, 76, 146, 149,
   150, 157
Touch stress   5, 33, 34, 46, 57
TPRK125   111
Transgenic plants   58, 59, 67, 74, 124, 125
Transmembrane domains   41, 47, 151, 155
Triple response   8, 146, 147
Tubulin   107, 109

*Verticillium dahliae*   78
*Vicia faba*   138
VirA receptor   150

WCK1   15, 18
Wee1   101
Wheat   14, 42, 44
WIN1   46
WIPK   6, 15, 19, 33, 34, 47, 55–59, 70–81, 88,
   152
Wound-inducible genes   53
Wounding   5, 6, 19, 20, 33, 34, 43, 46, 47,
   53–60, 70, 76–78, 80, 102, 106, 107, 131, 145,
   152

*Xenopus*   103, 109

Yeast 2-hybrid   4, 34, 60, 156

# Springer
# and the
# environment

At Springer we firmly believe that an international science publisher has a special obligation to the environment, and our corporate policies consistently reflect this conviction.
We also expect our business partners – paper mills, printers, packaging manufacturers, etc. – to commit themselves to using materials and production processes that do not harm the environment. The paper in this book is made from low- or no-chlorine pulp and is acid free, in conformance with international standards for paper permanency.

 Springer

Printing (Computer to Film): Saladruck, Berlin
Binding: Stürtz AG, Würzburg